OXFORD MATHEMATICAL MONOGRAPHS

OXFORD MATHEMATICAL MONOGRAPHS

THE LAMINAR BOUNDARY LAYER EQUATIONS

By N. CURLE. 1962

MEROMORPHIC FUNCTIONS

By W. K. HAYMAN. 1963

THE THEORY OF LAMINAR BOUNDARY LAYERS IN COMPRESSIBLE FLUIDS

By K. STEWARTSON. 1964

HOMOGRAPHIES, QUATERNIONS AND ROTATIONS

By P. DU VAL. 1964

THE OPEN MAPPING AND CLOSED GRAPH THEOREMS IN TOPOLOGICAL VECTOR SPACES

By T. HUSAIN. 1965

MATHEMATICAL THEORY OF CREEP AND CREEP RUPTURE

By F. K. G. ODQVIST. 1966

CLASSICAL HARMONIC ANALYSIS AND LOCALLY COMPACT GROUPS

By H. REITER. 1968

QUANTUM-STATISTICAL FOUNDATIONS OF CHEMICAL KINETICS

By S. GOLDEN. 1969

VARIATIONAL PRINCIPLES IN HEAT TRANSFER

A Unified Lagrangian Analysis of Dissipative Phenomena

BY

MAURICE A. BIOT

Member U.S. National Academy of Engineering
Foreign Member Royal Academy of Belgium

OXFORD
AT THE CLARENDON PRESS
1970

Oxford University Press, Ely House, London W. 1

GLASGOW NEW YORK TORONTO MELBOURNE WELLINGTON
CAPE TOWN SALISBURY IBADAN NAIROBI DAR ES SALAAM LUSAKA ADDIS ABABA
BOMBAY CALCUTTA MADRAS KARACHI LAHORE DACCA
KUALA LUMPUR SINGAPORE HONG KONG TOKYO

PRINTED IN GREAT BRITAIN

PREFACE

THIS book develops a variational treatment of heat transfer which includes heat conduction and convection. It is intended to provide the foundation of a unified analysis of irreversible processes by methods analogous to those of classical mechanics. This opens the way to a formulation of heat transfer and dissipative phenomena in complex systems by means of Lagrangian-type equations and generalized coordinates. At the same time this approach suggests many approximate procedures and drastic simplifications applicable to the practical solution of a large category of problems in physics and technology.

The variational analysis referred to here must be understood to extend beyond the more restricted traditional viewpoint, based on the condition that the value of a given scalar is stationary. The methods involve a generalization of the concept of virtual work in classical mechanics. Furthermore, the principle of virtual work itself is brought into a still more general perspective derived from the concept of 'variational scalar product'. As a consequence variational equations are obtained containing terms of various types, some of which are true variations of scalars while others are similar to those defining the generalized forces in Lagrangian mechanics.

The main body of the book, which contains eight chapters, treats the subject exclusively in the context of heat transfer. This provides a convenient background in order to illustrate diversified viewpoints and procedures that may be used in many practical problems. Applications to other areas of physics are discussed in an appendix. Some of the more fundamental mathematical aspects of the subject, from the viewpoint of functional analysis, are also discussed in the appendix.

Chapter 1 introduces in the context of linear heat conduction a fundamental form of variational principle based on the concepts of thermal potential, dissipation function, and generalized thermal force. Conservation of energy plays the role of a holonomic constraint. The dissipation function, expressed in terms of the time derivative of a heat-displacement vector, is a generalization of the concept introduced by Rayleigh for mechanical systems with viscous dissipation. This generalization is closely related to Onsager's principle of reciprocity in the thermodynamics of irreversible processes. The resulting Lagrangian equations, with generalized coordinates, lead to minimum dissipation principles. An application to the simple problem of heat penetration through a wall, using the concept of penetration depth, illustrates the power and accuracy of the method.

The general properties of linear systems and the corresponding linear Lagrangian equations are derived in Chapter 2. The physical system includes a local linear boundary heat-transfer condition, which is frequently used in practice as an approximation. This is accomplished by incorporating the boundary dissipation into the dissipation function describing the whole system. One of the important features of linear systems is the existence of relaxation modes and normal coordinates. An interesting aspect of normal coordinates in heat conduction is the property of infinite degeneracy associated with steady-state flow. The use of normal coordinates may lead to 'weak solutions' in the sense of functional analysis as illustrated by discussing again the problem of penetration of heat into a wall.

An operational formulation for the linear case with time-independent parameters is developed in Chapter 3. The general form of thermal admittances and impedances is established on the basis of the non-negative and positive-definite character of the fundamental quadratic forms describing the system. These results establish the thermal response for harmonic time-dependence. Transients are then obtained on the basis of Fourier–Laplace transforms. The resulting operational rules are considerably simplified by retaining for the time derivative the simple operator originally introduced by Heaviside. This amounts to a particular interpretation of the Laplace transform in terms of generalized functions. By their very nature the operational equations lead immediately to variational principles in operational form. They are open to various interpretations by considering the differential operator as an ordinary algebraic or numerical variable or in the time domain by interpreting the products of Laplace transforms as convolutions. A general rule for the interconnection of thermal systems is also derived in this variational context and provides a general basis for the formulation of a variety of finite element methods. An example also illustrates the concept of continuous relaxation spectrum and its relation to operational methods.

The method of associated fields presented in Chapter 4 constitutes a procedure by which the temperature field may be decoupled from the heat displacement vector field with zero divergence. The latter corresponds to ignorable coordinates in the sense of classical mechanics and may be eliminated from the problem when only the temperature is to be evaluated. The method is shown to be related to the existence of normal coordinates with infinite degeneracy. It is illustrated by an application to the problem of heating of a flange and web structure.

Extension of the variational principles to non-linear systems with temperature-dependent parameters is the object of Chapter 5. Generalizing the concept of thermal potential leads to equations similar to those obtained for linear systems. Under certain conditions the method of associated fields may also be used in non-linear problems. Particular problems with non-linear features such as radiation and melting boundaries are also discussed. The difference between heating and cooling due to non-linearity is brought out by treating numerically a simple case.

In Chapter 6 the variational methods are extended to convective heat transfer for linear and non-linear problems. Two types of approach are developed. The first of these considers the heat conduction in a fixed solid whose boundaries are in contact with a moving fluid. The heat transfer properties at the boundaries are embodied in a trailing function. This function is defined as the boundary temperature distribution due to a unit rate of heat injection into the fluid at a given point of the boundary, assuming this boundary to be adiabatic elsewhere. When this trailing function is known it is possible to formulate the over-all thermal problem, including the boundary convection, by Lagrangian equations. The method of associated fields is also extended to this case. A second approach is to treat the fixed solid and the moving fluid as a single unified system. Variational principles and Lagrangian equations are then established for the total system by using a definition of the dissipation function which includes convection.

The problem of boundary-layer heat transfer, which is treated in Chapter 7, is of considerable practical interest. The trailing function in this case may be evaluated by variational methods and provides a key representation of the heat transfer properties. The procedure is based on a conduction analogy whereby the method used for thermal conduction is readily applicable to boundary-layer convection. This provides an analysis of the problem that brings out the significant physical features and the basic difference between laminar and turbulent cases. They correspond to two distinct types of trailing functions of universal character. The method provides simple but physically sound procedures for the solution of many technological problems such as those of heat exchangers and aerodynamic heating.

Chapter 8 deals with complementary principles. Physical systems may usually be described by two types of variables denoted as intensive and extensive. In mechanics, for example, such a dual system of variables is represented by the forces and conjugate displacements. Variational principles formulated in terms of displacements lead to complementary principles in terms of forces. In the present case of purely thermal phenomena the temperature and the heat displacement may be considered respectively as intensive and extensive variables. In analogy with mechanics this leads to a duality whereby variational principles may be formulated as in the preceding chapters by means of heat displacements or in complementary form by means of the temperature. These complementary principles are derived for linear and non-linear systems with conduction and convection. An operational formulation along with interconnection and finite element methods are also discussed from the complementary viewpoint.

In the eight chapters of the book the analysis is presented in the restricted context of heat transfer. However, both the physical and mathematical contents of the subject are much broader. An appendix has therefore been added with the purpose of providing a more general perspective. Applications to other areas of physics are indicated, such as mass transport and the thermodynamics of irreversible processes. An illustration is provided by the Lagrangian analysis of thermoelasticity. An obvious application is to viscous fluids using Rayleigh's classical dissipation function. Electromagnetism may also be formulated by similar methods. From the purely mathematical viewpoint a broader perspective based on the concept of variational scalar product is also obtained. Essentially this concept provides a powerful tool for handling transformations in functional space. It implies such procedures as the transformation of linear into non-linear differential equations using coordinates analogous to a penetration depth. These considerations bring into a unified framework methods known under a variety of names in applied mathematics. In addition, because of the existence of a resolution threshold in physical problems, it is possible to consider a more realistic definition of the notion of completeness for generalized coordinates that takes into account the discrete particle nature of matter, in contrast with the continuous mathematical model.

The preparation of this book and part of the underlying original research were supported by the Air Force Office of Scientific Research under contract AF 49(638)–1329. During its terminal phase the work was sponsored through the European Office of Aerospace Research, OAR, United States Air Force under contract F 61052–69–C–0030.

M. A. B.

Brussels
June 1969

CONTENTS

CHAPTER ONE

FUNDAMENTAL VARIATIONAL PRINCIPLE IN THERMAL CONDUCTION

1. INTRODUCTION

In order to familiarize the reader with the fundamental concepts and methods developed in this book we shall limit ourselves in this first chapter to thermal conduction in a system with properties independent of the temperature. Such a system is physically linear. The more general non-linear case for a medium with properties dependent on the temperature will be considered in Chapter 5.

The key concept of heat displacement field is introduced in Section 2. This concept leads to a fundamental variational principle which is first developed in the context of isotropic conductivity. There are two essential features involved here which follow naturally from the concept of heat displacement. One is that the temperature and the heat displacement are conjugate variables analogous to force and displacement in mechanics. The other is that conservation of energy is verified identically by the choice of variables describing the system in a way similar to a holonomic constraint in mechanics. One of the advantages of this approach lies in the particular form of the associated variational principle by which it is possible to verify approximately the law of heat conduction while maintaining exact energy conservation.

An important feature of this method also results from the possibility of extending to thermodynamics the *principle of virtual work* and the concept of generalized force.

The variational principle thus obtained may be termed fundamental in contrast to a complementary form. The existence of these two forms follows from a similar duality in classical mechanics where variational principles are expressed in a fundamental form in terms of displacements or a complementary form in terms of forces. These complementary principles will be discussed in Chapter 8.

Another significant advantage of the fundamental form of the variational principle is the absence of any space derivative of the temperature in its formulation. This results in higher accuracy in applications to approximate solutions. More flexibility is also obtained in the choice of such solutions since discontinuities may be introduced in the approximate representation of the temperature field. The temperature gradient itself does not have to be closely matched.

The concept of generalized coordinates, together with the variational principle, leads to Lagrangian-type differential equations by means of expressions analogous to mechanical forces and potentials and Rayleigh's dissipation function. A principle of minimum dissipation is also obtained as a corollary. The fundamental significance of the generalized coordinates as providing a complete description of the physical system is discussed in the context of discrete molecular structure.

These results are derived in Sections 3 and 4. In Section 5 they are extended to the case of anisotropic thermal conductivity. The definition of a dissipation function for this case requires the application of the thermodynamics of irreversible processes based on Onsager's relations. The addition of continuously distributed heat sources is considered in Section 6 and the corresponding Lagrangian equations are derived for this case.

In the last section the results are applied to the simple problem of the heating of a slab, which provides a good illustration of both the accuracy and flexibility of this approach. In particular, it is shown how great simplification is achieved by using the concept of penetration depth as a generalized coordinate in the initial phase of the phenomenon. Although the system is physically linear, the initial phase is thus governed by a non-linear equation, while the second phase is described by different coordinates that lead to linear equations.

The principles presented in this chapter in the context of heat conduction constitute a particular case of a general approach to linear irreversible thermodynamics developed by the author in 1954.‡ It was shown that a large class of irreversible processes may be described by generalized coordinates obeying the same Lagrangian equations as in the classical mechanics of dissipative systems. These results are applicable to a thermodynamic system with non-uniform temperature. This was illustrated in a second paper§ dealing with the special case of coupled thermoelasticity. The analysis of linear thermal conduction is obtained

‡ M. A. Biot, 'Theory of stress-strain relations in anisotropic viscoelasticity and relaxation phenomena', *J. appl. Phys.* **25**, 1385–91 (1954).

A more extensive account of the general theory will be found in a later paper: M. A. Biot, 'Linear thermodynamics and the mechanics of solids', *Proceedings of the Third U.S. National Congress of Applied Mechanics*, pp. 1–18. American Society of Mechanical Engineers, New York (1958).

§ M. A. Biot, 'Thermoelasticity and irreversible thermodynamics', *J. appl. Phys.* **27**, 240–53 (1956).

as a particular case of the more general thermodynamics by restricting the phenomenon to purely thermal exchanges. This application to thermal conduction was developed in detail in a third paper.‡ The analysis presented in this chapter is based essentially on the material contained in this third paper.

2. VARIATIONAL PRINCIPLE FOR ISOTROPIC THERMAL CONDUCTIVITY

We consider a solid of isotropic thermal properties independent of the temperature. The medium may be homogeneous or non-homogeneous. In the latter case the thermal conductivity $k(x,y,z)$ and the heat capacity $c(x,y,z)$ *per unit volume* are functions of the coordinates x, y, z. The heat capacity per unit volume, rather than its usual definition per unit mass, is introduced in order to avoid the use of the symbol ρ for the density, since it does not play any role in the particular category of phenomena considered throughout this book.

The classical description of thermal phenomena is based on the temperature as a scalar field. In the present variational procedure, an essential feature results from the introduction of a vector field into the basic laws of heat conduction. This vector field, $\mathbf{H}(x,y,z,t)$, which we shall call the *heat displacement*, is a function of the time and the coordinates. It is defined by the equation

$$\dot{\mathbf{H}} = \frac{\partial}{\partial t}\mathbf{H}(x,y,z,t), \tag{2.1}$$

where $\dot{\mathbf{H}}$ is the vector representing the local rate of heat flow per unit area. Hence the vector $\mathbf{H}$ is the time integral of the rate of flow vector $\dot{\mathbf{H}}$.

The reason for using this particular description of the heat flow becomes apparent when we consider the law of conservation of energy

$$c\theta = -\operatorname{div}\mathbf{H}. \tag{2.2}$$

In this form it does not involve any time derivative and may be considered as a *holonomic constraint* in the sense of classical mechanics. Hence equation (2.2) must be looked on as something more than a relation to be verified by a physical solution. It must also be verified by variations $\delta\mathbf{H}$ and $\delta\theta$ in analogy with virtual displacements. Therefore we may write

$$c\,\delta\theta = -\operatorname{div}(\delta\mathbf{H}). \tag{2.3}$$

‡ M. A. Biot, 'New methods in heat flow analysis with application to flight structures', *J. aeronaut. Sci.* **24**, 857–73 (1957).

By the foregoing procedure, use of the time derivative is relegated to the law of heat conduction, which is written as

$$\operatorname{grad}\theta+\frac{1}{k}\dot{\mathbf{H}}=0. \tag{2.4}$$

Equations (2.2) and (2.4) provide a complete formulation of heat conduction under the present assumptions.

Variational principles are obtained as follows. Consider a variation $\delta\mathbf{H}$ of the field $\mathbf{H}$ and corresponding variations $\delta\theta$ given by the constraint condition (2.3). We multiply equation (2.4) by $\delta\mathbf{H}$ and integrate over a volume τ of the medium. We obtain

$$\iiint_{\tau}\left(\operatorname{grad}\theta+\frac{1}{k}\dot{\mathbf{H}}\right).\delta\mathbf{H}\,d\tau=0. \tag{2.5}$$

Integration by parts of the first term yields

$$\iiint_{\tau}\left\{-\theta\operatorname{div}(\delta\mathbf{H})+\frac{1}{k}\dot{\mathbf{H}}.\delta\mathbf{H}\right\}d\tau=-\iint_{A}\theta\mathbf{n}.\delta\mathbf{H}\,dA, \tag{2.6}$$

where $\mathbf{n}$ is the unit normal pointing outward at the boundary surface A. From equation (2.3) we derive

$$-\iiint_{\tau}\theta\operatorname{div}(\delta\mathbf{H})\,d\tau=\iiint_{\tau}c\theta\,\delta\theta\,d\tau=\delta V. \tag{2.7}$$

The scalar V is defined as

$$V=\tfrac{1}{2}\iiint_{\tau}c\theta^{2}\,d\tau. \tag{2.8}$$

It plays the role of a potential, as will be made clearer by the discussion of this quantity in connection with generalized coordinates in Section 3 below. Introduction of δV into equation (2.6) yields

$$\delta V+\iiint_{\tau}\frac{1}{k}\dot{\mathbf{H}}.\delta\mathbf{H}\,d\tau=-\iint_{A}\theta\mathbf{n}.\delta\mathbf{H}\,dA. \tag{2.9}$$

This result expresses the variational principle for thermal conduction. Equation (2.9) must be verified for arbitrary variations of the field $\mathbf{H}$, with θ defined as a function of $\mathbf{H}$ by the constraint (2.2). Hence the variational principle (2.9) is nothing but a *statement of the heat conduction law* (2.4) with energy conservation verified automatically as a constraint.

The volume integral of $(1/k)\dot{\mathbf{H}}.\delta\mathbf{H}$ has an important physical meaning associated with the concept of dissipation, as will be shown in the next section by introducing generalized coordinates.

Time-dependent conductivity

The foregoing derivation of the variational principle remains valid if the thermal conductivity is not only a function of the coordinates but is also time-dependent, namely, if

$$k = k(x, y, z, t). \tag{2.10}$$

The usefulness of this remark results not so much from the actual physical occurrence of such a property but rather in connection with other phenomena leading to equations mathematically equivalent to the case of a time-dependent conductivity.

Moving boundaries

The variational principle (2.9) is also applicable to moving boundaries. This can be seen by noting that the volume and surface integrals are extended to instantaneous geometric configurations that do not have to be the same at different instants. Hence, the variational principle is nothing but a formulation of the law that governs the distribution of the heat flow rate $\dot{\mathbf{H}}$ at any particular instant for a given distribution of temperature θ. The variational principle may therefore be used to solve problems of heat conduction with melting boundaries.

Fourier's equation

In the present analysis the thermal field is governed by two equations. One of these, equation (2.2), expresses conservation of energy, the other, equation (2.4), expresses the law of heat conduction. This separation is a direct consequence of the introduction of the heat displacement field $\mathbf{H}$ as an additional variable. By eliminating $\mathbf{H}$ between equations (2.2) and (2.4) we obtain

$$\operatorname{div}(k \operatorname{grad} \theta) = c\frac{\partial \theta}{\partial t}. \tag{2.10 a}$$

This is the well-known Fourier equation commonly used to describe heat diffusion. It combines conservation of energy and the conduction law into a single equation. However, in many cases it is preferable to use the two separate equations, (2.2) and (2.4). In particular, this separation leads to the variational principle (2.5) by which it is possible to verify approximately the heat conduction law while maintaining exact heat conservation.

Generalized variational methods

The variational principle as expressed by equation (2.9) must be understood in a broader sense that extends beyond the traditional concepts of the variational calculus. Actually, this broader viewpoint is not new and corresponds to what is known in classical mechanics as *the principle of virtual work*. In dynamics it is also formulated as d'Alembert's principle. These principles lead to variational equations completely analogous to equation (2.9). From a purely abstract viewpoint the method is based on a mathematical concept that may be called the *variational scalar product*. It is an expression that represents a scalar product in functional space. In this expression some of the terms may be transformed into quantities that represent variations of invariants or functionals and correspond to the viewpoint of the traditional variational calculus. The remaining terms, which are not expressible in this way, fall in the category of what is designated in Lagrangian mechanics as the generalized forces. These broader aspects of variational methods are discussed in more detail in Section 4 of the Appendix and it is indicated how they may be translated into the more abstract formalism of functional analysis and the theory of sets.

3. GENERALIZED COORDINATES

The variational principle (2.9) acquires a new meaning if we express it in terms of generalized coordinates. Let the field $\mathbf{H}$ be represented in the form

$$\mathbf{H} = \mathbf{H}(q_1, q_2, \ldots, q_n, x, y, z, t). \tag{3.1}$$

In other words, we assume the field to be a *given function* of the space coordinates x, y, z, of the time t, and of a certain number of parameters $q_1, q_2, \ldots, q_n$. These parameters are unknown functions of time, which may be considered as *generalized coordinates* representing the field $\mathbf{H}$. We may use a finite number of coordinates q_i or an infinite denumerable set, depending on the type of problem considered and the accuracy required in the representation of the unknown field. In particular, we may choose a linear representation in the form of a finite or infinite series. In this case we write

$$\mathbf{H} = \sum^{i} q_i \mathbf{H}_i(x, y, z, t) \tag{3.2}$$

or

$$\mathbf{H} = \sum^{i} q_i \mathbf{H}_i(x, y, z), \tag{3.3}$$

where $\mathbf{H}_i$ is a field which is either a given function of time or independent of time.

Expressions (3.3) include as a particular case the Fourier series and expansions in orthogonal functions. The representation of the field $\mathbf{H}$ by generalized coordinates must therefore be understood as valid *almost everywhere*. In the mathematical language of abstract spaces this means that the representation is valid except for a set of points either finite or of zero measure.

The variational principle (2.9) may be formulated by means of the generalized coordinates instead of the components of the field itself. Consider arbitrary variations δq_i of the generalized coordinates. The corresponding field variations are

$$\delta \mathbf{H} = \sum^{i} \frac{\partial \mathbf{H}}{\partial q_i} \delta q_i. \tag{3.4}$$

Expression (2.8) for V is a function of q_i and we may write

$$\delta V = \sum^{i} \frac{\partial V}{\partial q_i} \delta q_i. \tag{3.5}$$

By taking into account relations (3.4) and (3.5), equation (2.9) is written

$$\sum^{i} \left(\frac{\partial V}{\partial q_i} + \iiint_{\tau} \frac{1}{k} \dot{\mathbf{H}} . \frac{\partial \mathbf{H}}{\partial q_i} \, d\tau \right) \delta q_i = - \sum^{i} \delta q_i \iint_{A} \theta \frac{\partial \mathbf{H}}{\partial q_i} . \mathbf{n} \, dA. \tag{3.6}$$

Since the variations δq_i are arbitrary, this last result implies the relations

$$\frac{\partial V}{\partial q_i} + \iiint_{\tau} \frac{1}{k} \dot{\mathbf{H}} . \frac{\partial \mathbf{H}}{\partial q_i} \, d\tau = - \iint_{A} \theta \frac{\partial \mathbf{H}}{\partial q_i} . \mathbf{n} \, dA. \tag{3.7}$$

We may write as many such equations as there are generalized coordinates.

Completeness of the generalized coordinates as a physical representation

It is always possible to choose the generalized coordinates in such number and in such a way that from the physical standpoint the system is completely described by these coordinates. This becomes evident when we remember that we are dealing with molecular structures and that the continuum is substituted for its description as an approximation.

Hence physically it is just as valid to consider a discrete system made up of a large number of finite cells. The cells may be very small but still large enough with respect to the molecular scale so that statistical thermodynamic laws remain valid for each cell. We may visualize cubic cells with a heat displacement vector attached to each face. The temperature of the cell is then determined in terms of the heat displacement on each face by a conservation law corresponding to equation (2.2). The discrete vectors thus defined completely describe the physical system and may themselves be considered as a special case of generalized coordinates.

4. LAGRANGIAN EQUATIONS AND MINIMUM DISSIPATION

The volume integral in equations (3.7) may be expressed in a simpler form that brings out its physical significance. Since the generalized coordinates q_i are functions of time, we may write

$$\dot{\mathbf{H}} = \sum^{i} \frac{\partial \mathbf{H}}{\partial q_i} \dot{q}_i + \frac{\partial \mathbf{H}}{\partial t}. \tag{4.1}$$

This result is obtained by taking the derivative of expression (3.1) with respect to time. From equation (4.1) we derive

$$\frac{\partial \dot{\mathbf{H}}}{\partial \dot{q}_i} = \frac{\partial \mathbf{H}}{\partial q_i}. \tag{4.2}$$

Hence, by putting

$$D = \frac{1}{2} \iiint_{\tau} \frac{1}{k} \dot{\mathbf{H}}^2 \, d\tau, \tag{4.3}$$

we find

$$\frac{\partial D}{\partial \dot{q}_i} = \iiint_{\tau} \frac{1}{k} \dot{\mathbf{H}} \cdot \frac{\partial \dot{\mathbf{H}}}{\partial \dot{q}_i} \, d\tau = \iiint_{\tau} \frac{1}{k} \dot{\mathbf{H}} \cdot \frac{\partial \mathbf{H}}{\partial q_i} \, d\tau. \tag{4.4}$$

We also put

$$Q_i = - \iint_{A} \theta \frac{\partial \mathbf{H}}{\partial q_i} \cdot \mathbf{n} \, dA. \tag{4.5}$$

After substituting the values (4.4) and (4.5) into equations (3.7) the latter become

$$\frac{\partial V}{\partial q_i} + \frac{\partial D}{\partial \dot{q}_i} = Q_i. \tag{4.6}$$

If there are n generalized coordinates q_i, these equations constitute a system of n differential equations for the unknowns q_i.

Equations (4.6) are of the same form as those of Lagrangian mechanics for the slow motion of a dissipative system with negligible inertia forces. The function V is the potential energy, and D is the dissipation function. On the right-hand side, Q_i is a generalized external force applied to the system and defined by a method of virtual work.

Similarly, in a thermal conduction problem we shall refer to V as a *thermal potential* and to D as a *dissipation function*.

The quantities Q_i represent generalized thermal driving forces due to the temperature distribution at the boundary. For this reason they are referred to as *thermal forces*. They may be defined by a method of virtual work as in mechanics. This can be seen by considering the following relation obtained from equations (3.4) and (4.5):

$$\sum^{i} Q_i \delta q_i = -\sum^{i} \iint_A \theta \frac{\partial \mathbf{H}}{\partial q_i} . \mathbf{n}\, \delta q_i \, dA = -\iint_A \theta \mathbf{n} . \delta \mathbf{H} \, dA. \quad (4.7)$$

Assume that we vary only one coordinate q_i by a unit amount δq_i. The right side of equation (4.7) then represents the corresponding thermal force Q_i. In terms of a mechanical model expression (4.7) may be interpreted as the virtual work done by a pressure θ on a volume inflow of fluid $-\mathbf{n}.\delta\mathbf{H}$ per unit area at the boundary.

Minimum dissipation principle

The particular form of the Lagrangian equations (4.6) immediately suggests that they are equivalent to a minimum dissipation principle. This was shown in 1955 by the author in the more general context of irreversible thermodynamics,‡ which includes heat conduction as a special case.

We introduce the quantity

$$X_i = Q_i - \frac{\partial V}{\partial q_i}. \quad (4.8)$$

The Lagrangian equations (4.6) become

$$\frac{\partial D}{\partial \dot{q}_i} = X_i. \quad (4.9)$$

Let us now assume that $\mathbf{H}$ does not contain the time explicitly; the value (4.1) of $\dot{\mathbf{H}}$ is then

$$\dot{\mathbf{H}} = \sum^{i} \frac{\partial \mathbf{H}}{\partial q_i} \dot{q}_i \quad (4.10)$$

and the dissipation function D is a positive quadratic form in the variables $\dot{q}_i$. In that case equations (4.9) show that $X_i = 0$ implies $\dot{q}_i = 0$, which is a condition of static equilibrium. Hence we have called X_i the *disequilibrium forces*.

For given values of q_i the dissipation function depends on the time derivatives $\dot{q}_i$. Let us impose the condition that D be a minimum when the vector $\dot{q}_i$ is varied subject to the constraint

$$\sum^{i} X_i \dot{q}_i = \text{const.} \quad (4.11)$$

This is equivalent to the absolute variational condition

$$\sum^{i} \left(\frac{\partial D}{\partial \dot{q}_i} - \Lambda X_i \right) \delta \dot{q}_i = 0, \quad (4.12)$$

‡ M. A. Biot, 'Variational principles in irreversible thermodynamics with application to viscoelasticity', *Phys. Rev.* **97**, 1463–9 (1955).

where Λ is an undetermined Lagrangian multiplier. By putting $\Lambda = 1$ the variational condition (4.12) leads to equations (4.9). Hence the minimum dissipation principle under the constraint (4.11) determines the time rate of change of the thermal system at any given instant. This provides a physical interpretation of the Lagrangian equations.

Note that in a mechanical or electrical analogue model the constraint (4.11) expresses that the power input of the disequilibrium forces is constant.

An equivalent variational principle is also obtained by stating that the expression

$$P = D - \sum^{i} X_i \dot{q}_i \tag{4.13}$$

be a minimum when varying $\dot{q}_i$, since it implies the same equations (4.9).

As readily verified, the foregoing variational principles are valid also if $\mathbf{H}$ contains the time explicitly. However, the physical interpretation becomes somewhat more involved since $\dot{q}_i = 0$ does not correspond to static equilibrium in this case.

A closely related form of these variational principles is also obtained by going back to the heat conduction law (2.4), which is written

$$\operatorname{grad} \theta + \frac{1}{k} \dot{\mathbf{H}} = 0. \tag{4.14}$$

Consider the expression

$$P = \iiint_{\tau} \left(\dot{\mathbf{H}} . \operatorname{grad} \theta + \frac{1}{2k} \dot{\mathbf{H}}^2 \right) d\tau, \tag{4.15}$$

where τ is the volume of the thermal system. The condition that P is a minimum when $\dot{\mathbf{H}}$ is varied leads to the heat conduction equation (4.14). This is analogous to minimizing the previous expression (4.13) for P. Similarly, we verify that the heat conduction law is expressed by the condition that

$$D = \frac{1}{2} \iiint_{\tau} \frac{1}{k} \dot{\mathbf{H}}^2 \, d\tau \tag{4.16}$$

be a minimum under the constraint

$$\iiint_{\tau} \dot{\mathbf{H}} . \operatorname{grad} \theta \, d\tau = \text{const.} \tag{4.17}$$

Comparing with equation (4.11), we see that $\operatorname{grad} \theta$ plays the role of a local disequilibrium force. These results are readily extended to anisotropic conductivity.

Fundamental significance of the Lagrangian equations

As pointed out in the previous section, it is theoretically possible to choose a finite but large number of generalized coordinates in such a way that the system is described completely from the physical standpoint, provided we remain within a '*resolution threshold*'. This concept is discussed in more detail in Section 4 of the Appendix. In this sense the Lagrangian equations provide an accurate formulation of the physical behaviour. They also have the particular advantage of bringing to light the fundamental mathematical structure of a problem. In addition, this is accomplished in a way that is independent not only of the frame of reference but also of any particular representation of the unknowns.

Lagrangian equations as a method of approximate analysis

In numerous problems it is possible to guess that a solution belongs to a family of functions with one or more unknown parameters. For example, we may guess that an unknown time-dependent field belongs to the family of functions

$$\mathbf{H} = \mathbf{H}(q_1, q_2, x, y, z), \tag{4.18}$$

where q_1 and q_2 are unknown parameters, functions of time, which play the role of generalized coordinates. We may also say that the variations $\delta\mathbf{H}$ are restricted by a holonomic constraint with two degrees of freedom corresponding to two arbitrary variations δq_1, δq_2. The variational principle is still satisfied for these variations and provides two Lagrangian differential equations to determine the time dependence of q_1 and q_2. As occurs frequently in physics and technology, certain characteristic features of a solution are already known. This may be the result of intuitive understanding, past experience, or mathematical properties. When this is the case, the Lagrangian formulation provides a method of confining the formulation exclusively to the unknown features. This furnishes a powerful tool leading to great simplification and flexibility in the approximate analysis of complex practical problems. The implications of this approach are also discussed in Section 4 of the Appendix in the broader context of functional analysis.

5. ANISOTROPIC THERMAL CONDUCTIVITY

Up to now we have assumed the thermal conductivity to be isotropic. The case of anisotropic conductivity will now be considered. It is not convenient in this case to use the vector notation and we shall represent the heat displacement field **H** by its components

$$H_i = H_i(x, y, z, t). \tag{5.1}$$

Equations (2.2) and (2.3) remain the same and are now written as

$$c\theta = -\sum^{i} \frac{\partial H_i}{\partial x_i}, \tag{5.2}$$

$$c\,\delta\theta = -\sum^{i} \frac{\partial}{\partial x_i}(\delta H_i). \tag{5.3}$$

The coordinates x, y, z are denoted by x_i. For anisotropic conductivity the law of heat conduction becomes

$$\sum^{j} k_{ij} \frac{\partial \theta}{\partial x_j} + \dot{H}_i = 0, \tag{5.4}$$

where k_{ij} is the thermal conductivity tensor.

An important property of the thermal conductivity is expressed by the relation

$$k_{ij} = k_{ji}. \tag{5.5}$$

Hence k_{ij} is a symmetric tensor with six independent components. This property is also a consequence of Onsager's principle. Equations (5.5) may be considered as a particular case of Onsager's reciprocity relations. These properties have been verified experimentally and are generally valid in the absence of a strong magnetic field.

In order to derive a variational principle we must write equations (5.4) in a form analogous to equation (2.4). Hence we solve the three equations (5.4) for the three components of the temperature gradient. This yields

$$\frac{\partial \theta}{\partial x_i} + \sum^{j} \lambda_{ij} \dot{H}_j = 0. \tag{5.6}$$

The tensor components λ_{ij} are the elements of the inverse of the matrix $[k_{ij}]$, i.e.

$$[\lambda_{ij}] = [k_{ij}]^{-1}. \tag{5.7}$$

The tensor λ_{ij} represents a *thermal resistivity*. It is also symmetric. Hence

$$\lambda_{ij} = \lambda_{ji}. \tag{5.8}$$

We now multiply (5.6) by δH_i and, after summation for all three components, we integrate the result over the volume τ. We obtain

$$\iiint_{\tau} \left(\sum^{i} \frac{\partial \theta}{\partial x_i} \delta H_i + \sum^{ij} \lambda_{ij} \dot{H}_j \, \delta H_i \right) d\tau = 0. \tag{5.9}$$

After integration by parts, and taking into account equations (5.3) and (2.8), we obtain

$$\delta V + \iiint_{\tau} \left(\sum^{ij} \lambda_{ij} \dot{H}_j \, \delta H_i\right) d\tau = - \iint_{A} \sum^{i} \theta n_i \, \delta H_i \, dA, \tag{5.10}$$

where n_i denotes the components of the unit normal pointing outward at the boundary surface A and V is the thermal potential defined by equation (2.8).

Consider the heat displacement field to be described by n generalized coordinates q_k. The components of this field are written as

$$H_i = H_i(q_1, q_2, \ldots, q_n, x, y, z, t). \tag{5.11}$$

We derive

$$\dot{H}_i = \sum^{k} \frac{\partial H_i}{\partial q_k} \dot{q}_k + \frac{\partial H_i}{\partial t}; \tag{5.12}$$

hence

$$\frac{\partial H_i}{\partial q_k} = \frac{\partial \dot{H}_i}{\partial \dot{q}_k} \tag{5.13}$$

and

$$\delta H_i = \sum^{k} \frac{\partial H_i}{\partial q_k} \delta q_k = \sum^{k} \frac{\partial \dot{H}_i}{\partial \dot{q}_k} \delta q_k. \tag{5.14}$$

Consider now the volume integral in equation (5.10). By using expressions (5.14) the integrand becomes

$$\sum^{ij} \lambda_{ij} \dot{H}_j \, \delta H_i = \sum^{k} \sum^{ij} \lambda_{ij} \dot{H}_j \frac{\partial \dot{H}_i}{\partial \dot{q}_k} \delta q_k. \tag{5.15}$$

Here we must introduce for the first time the reciprocity property

$$\lambda_{ij} = \lambda_{ji}. \tag{5.16}$$

By taking this relation into account, equation (5.15) may be written as

$$\sum^{ij} \lambda_{ij} \dot{H}_j \, \delta H_i = \sum^{k} \delta q_k \frac{\partial}{\partial \dot{q}_k} \left(\tfrac{1}{2} \sum^{ij} \lambda_{ij} \dot{H}_i \dot{H}_j \right). \tag{5.17}$$

We put

$$D = \tfrac{1}{2} \iiint_{\tau} \sum^{ij} \lambda_{ij} \dot{H}_i \dot{H}_j d\tau. \tag{5.18}$$

This expression defines the dissipation function for anisotropic conductivity. With this definition the volume integral in equation (5.10) becomes

$$\iiint_{\tau} \sum^{ij} \lambda_{ij} \dot{H}_j \, \delta H_i \, d\tau = \sum^{k} \frac{\partial D}{\partial \dot{q}_k} \delta q_k. \tag{5.19}$$

On the other hand, we may write

$$\delta V = \sum^{k} \frac{\partial V}{\partial q_k} \delta q_k. \tag{5.20}$$

A thermal force Q_k may also be defined as in equation (4.7) by putting

$$\sum^{k} Q_k \, \delta q_k = -\iint \sum^{i} \theta n_i \, \delta H_i \, dA \tag{5.21}$$

with

$$Q_k = -\iint_A \sum^{i} \theta n_i \frac{\partial H_i}{\partial q_k} \, dA. \tag{5.22}$$

Since δq_k is arbitrary, substitution of the values (5.19), (5.20), and (5.21) into equation (5.10) yields

$$\frac{\partial V}{\partial q_k} + \frac{\partial D}{\partial \dot{q}_k} = Q_k. \tag{5.23}$$

Hence the Lagrangian equations (4.6) have been generalized to the case of anisotropic thermal conductivity.

The heat capacity per unit volume $c(x, y, z)$ may be a function of the coordinates. As in the case of isotropic conductivity discussed in Section 2, the results for the anisotropic case are also applicable for moving boundaries and for a thermal conductivity $k_{ij}(x, y, z, t)$ function of both the time and the coordinates.

6. HEAT SOURCES

We shall extend the results to a system containing heat sources. In order to simplify the writing, isotropic conductivity will be assumed. This can be done without loss of generality since the treatment is identical for the anisotropic case.

The rate of heat generation per unit volume and unit time is denoted by

$$w = w(x, y, z, t). \tag{6.1}$$

It is assumed to be a given function that may depend on time and location. The heat capacity $c(x, y, z)$ may be a function of the coordinates. Conservation of energy is expressed by the equation

$$c\theta = -\operatorname{div} \mathbf{H} + \int_0^t w \, dt. \tag{6.2}$$

Let the heat displacement $\mathbf{H}$ be represented by the superposition of two separate fields,

$$\mathbf{H} = \mathbf{H}^{+} + \mathbf{H}^{*}, \tag{6.3}$$

where $\mathbf{H}^*$ satisfies the equation

$$\operatorname{div} \mathbf{H}^* = \int_0^t w \, dt. \tag{6.4}$$

This equation does not determine $\mathbf{H}^*$ uniquely but we shall choose any particular field satisfying equation (6.4). Such a field may then be considered as a given function of the time and the coordinates. Substitution of $\mathbf{H}$ into equation (6.2) now yields

$$c\theta = -\operatorname{div}\mathbf{H}^{+}. \tag{6.5}$$

The unknown field $\mathbf{H}^{+}$ is assumed to be a function of the generalized coordinates

$$\mathbf{H}^{+} = \mathbf{H}^{+}(q_1, q_2,..., q_n, x, y, z, t). \tag{6.6}$$

We consider equation (6.5) to represent a holonomic constraint, valid for arbitrary variations of the field, hence

$$c\,\delta\theta = -\operatorname{div}\delta\mathbf{H}^{+}. \tag{6.7}$$

Actually, we vary only the generalized coordinates q_i, so that the variations $\delta\mathbf{H}^{+}$ and $\delta\mathbf{H}$ are the same, namely,

$$\delta\mathbf{H}^{+} = \delta\mathbf{H} = \sum^{i} \frac{\partial\mathbf{H}}{\partial q_i}\,\delta q_i. \tag{6.8}$$

Therefore, equation (6.7) may be written

$$c\,\delta\theta = -\operatorname{div}\delta\mathbf{H}. \tag{6.9}$$

As before, the law of heat conduction is

$$\operatorname{grad}\theta + \frac{1}{k}\dot{\mathbf{H}} = 0. \tag{6.10}$$

Equations (6.9) and (6.10) are identical to equations (2.3) and (2.4). In addition, as before, we may write

$$\frac{\partial\mathbf{H}}{\partial q_i} = \frac{\partial\dot{\mathbf{H}}}{\partial\dot{q}_i}. \tag{6.11}$$

This is the same as equation (4.2). Therefore the same procedure as in Sections 3 and 4 is applicable, leading to the Lagrangian equations

$$\frac{\partial V}{\partial q_i} + \frac{\partial D}{\partial\dot{q}_i} = Q_i. \tag{6.12}$$

The thermal potential V is a function of the unknown field $\mathbf{H}^{+}$ through θ, by equation (6.5). Let us examine the particular form of the dissipation function in this case. For isotropic thermal conductivity its value is

$$D = \frac{1}{2}\iiint_{\tau} \frac{1}{k}\dot{\mathbf{H}}^2\,d\tau, \tag{6.13}$$

with

$$\dot{\mathbf{H}} = \dot{\mathbf{H}}^{+} + \dot{\mathbf{H}}^{*} \tag{6.14}$$

and

$$\dot{\mathbf{H}}^{+} = \sum^{i} \frac{\partial\mathbf{H}^{+}}{\partial q_i}\dot{q}_i + \frac{\partial\mathbf{H}^{+}}{\partial t}. \tag{6.15}$$

We derive the following result:

$$\frac{\partial D}{\partial \dot{q}_i} = \frac{\partial D^+}{\partial \dot{q}_i} - Q_i^+, \tag{6.16}$$

where

$$D^+ = \frac{1}{2} \iiint_\tau \frac{1}{k} (\dot{\mathbf{H}}^+)^2 \, d\tau, \tag{6.17}$$

$$Q_i^+ = -\iiint_\tau \frac{1}{k} \frac{\partial \mathbf{H}^+}{\partial q_i} . \dot{\mathbf{H}}^* \, d\tau. \tag{6.18}$$

With these definitions the Lagrangian equations (6.12) take the form

$$\frac{\partial V}{\partial q_i} + \frac{\partial D^+}{\partial \dot{q}_i} = Q_i + Q_i^+. \tag{6.19}$$

The expression D^+ is the dissipation function due to the field $\mathbf{H}^+$. This field also determines θ, by equation (6.5). The term Q_i^+ represents a *generalized driving force due to the heat sources.*

The same results are obtained for the more general case of anisotropic conductivity by using expression (5.18) for the dissipation function.

Classical equations for heat conduction with thermal sources

The foregoing results have been derived by considering separately the two equations (6.2) and (6.10). Elimination of the field $\mathbf{H}$ between these two equations yields

$$c \frac{\partial \theta}{\partial t} = \operatorname{div}(k \operatorname{grad} \theta) + w. \tag{6.19 a}$$

This is the classical equation for the temperature field in the presence of heat sources. A similar equation for anisotropic thermal conductivity is obtained by using equation (5.4) instead of (6.10):

$$c \frac{\partial \theta}{\partial t} = \sum^{ij} \frac{\partial}{\partial x_i}\left(k_{ij} \frac{\partial \theta}{\partial x_j}\right) + w. \tag{6.19 b}$$

7. NUMERICAL EXAMPLE

Application of the foregoing results is well illustrated by the following problem, which brings to light some of the characteristic features of the method.

Consider a plate of thickness l with constant values of the thermal conductivity k and heat capacity c. The plate is initially at the temperature $\theta = 0$. At $t = 0$ one face of the plate, located at $x = 0$, is suddenly brought to a constant temperature $\theta = \theta_0$. The other face at $x = l$ is thermally insulated.

The heating process is divided into two phases. In the first phase it is assumed that the heat has penetrated to a depth $x = q_1$ smaller than the thickness l and that the temperature distribution is well approximated by the expression

$$\theta = \theta_0\left(1-\frac{x}{q_1}\right)^2. \tag{7.1}$$

This parabolic approximation is shown by curve (1) in Fig. 1.1. The *penetration depth* q_1 is a generalized coordinate to be determined as a function of time.

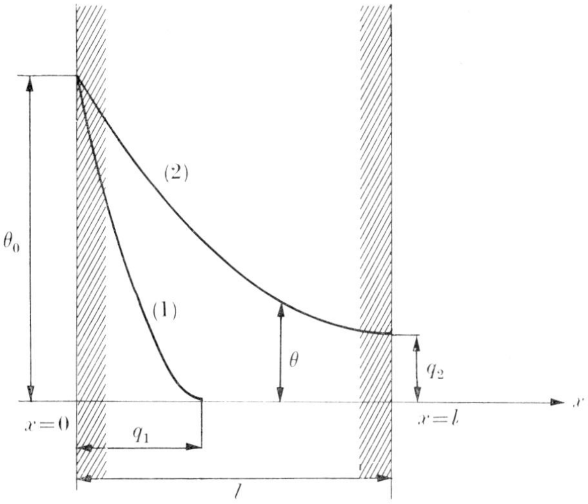

FIG. 1.1. Distribution of θ for first phase (1) and second phase (2) in heating of a slab thermally insulated at $x = l$.

Since this is a one-dimensional problem it is sufficient to consider a cylinder of unit cross-section of axis perpendicular to the wall.

The thermal potential is

$$V = \tfrac{1}{2}c\int_0^{q_1}\theta^2\,dx = \tfrac{1}{10}c\theta_0^2 q_1. \tag{7.2}$$

The heat displacement is derived from the temperature θ by using equation (2.2), which in this case becomes

$$c\theta = -\frac{dH}{dx}. \tag{7.3}$$

By taking into account the condition $H = 0$ at $x = q_1$; we derive

$$H = c\theta_0\left(\frac{q_1}{3}-x+\frac{x^2}{q_1}-\frac{x^3}{3q_1^2}\right). \tag{7.4}$$

The dissipation function is

$$D = \frac{1}{2k}\int_0^{q_1} \dot{H}^2\,dx = \frac{13}{630}\frac{c^2\theta_0^2}{k}q_1\dot{q}_1^2. \tag{7.5}$$

The generalized thermal force Q_1 is obtained by considering the virtual heat displacement $\delta H = \frac{1}{3}c\theta_0\,\delta q_1$ at $x = 0$. Applying equation (4.7), we write

$$Q_1\,\delta q_1 = \theta_0\,\delta H, \tag{7.6}$$

hence

$$Q_1 = \tfrac{1}{3}c\theta_0^2. \tag{7.7}$$

Equation (4.6) for the unknown Lagrangian coordinate q_1 is

$$\frac{\partial V}{\partial q_1}+\frac{\partial D}{\partial \dot{q}_1} = Q_1. \tag{7.8}$$

Substitution of the values (7.2), (7.5), and (7.7) yields

$$\frac{13}{315}q_1\dot{q}_1 = \frac{7}{30}\frac{k}{c}. \tag{7.9}$$

This is a first-order equation with the time differential $\dot{q}_1$. With the initial condition $q_1 = 0$ at $t = 0$ we find

$$q_1 = 3{\cdot}36\sqrt{\left(\frac{k}{c}t\right)}. \tag{7.10}$$

The first phase ends when $q_1 = l$ at a time t equal to

$$t_1 = 0{\cdot}0885\frac{cl^2}{k}. \tag{7.11}$$

This *transit time* measures the time required for heat to penetrate through a thickness l of a given material.

In the second phase, corresponding to times $t > t_1$, the temperature rises at the insulated boundary $x = l$. The temperature in this phase is also assumed to be well represented by a parabolic approximation,

$$\theta = (\theta_0 - q_2)\left(1-\frac{x}{l}\right)^2 + q_2. \tag{7.12}$$

This is illustrated by curve (2) in Fig. 1.1. The generalized coordinate q_2 is the unknown temperature at the boundary $x = l$. With the value (7.12) of θ the thermal potential is

$$V = \tfrac{1}{2}c\int_0^l \theta^2\,dx = (\tfrac{1}{10}\theta_0^2+\tfrac{2}{15}\theta_0 q_2+\tfrac{4}{15}q_2^2)cl. \tag{7.13}$$

The heat displacement H is obtained by integrating equation (7.3) with the value (7.12) of θ and the boundary condition $H = 0$ for $x = l$. We derive the dissipation function

$$D = \frac{1}{2k}\int\limits_0^l \dot{H}^2\,dx = \frac{34}{315}\frac{c^2l^3}{k}\dot{q}_2^2. \tag{7.14}$$

The generalized thermal force Q_2 is obtained by considering the virtual heat displacement $\delta H = \frac{2}{3}lc\,\delta q_2$. Applying equation (4.7), we write

$$Q_2\,\delta q_2 = \theta_0\,\delta H; \tag{7.15}$$

hence

$$Q_2 = \tfrac{2}{3}\theta_0\,lc. \tag{7.16}$$

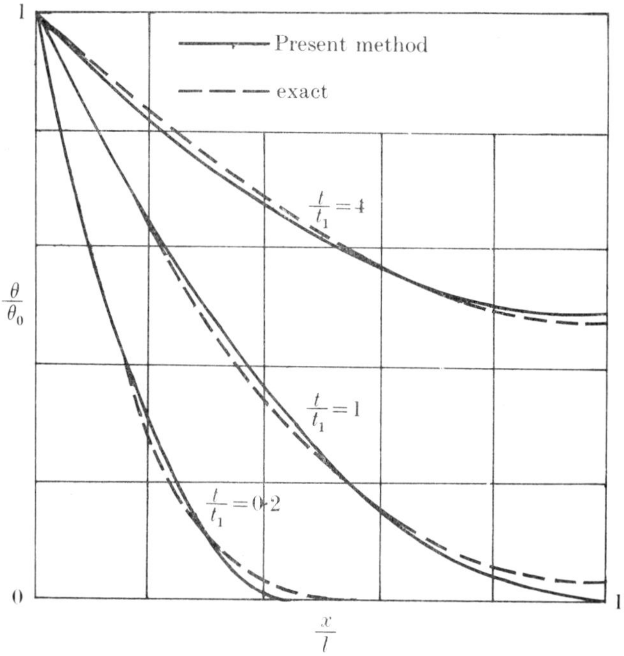

FIG. 1.2. Temperature distribution at various times in a slab.

By introducing the values (7.13), (7.14), and (7.16) in the Lagrangian equation

$$\frac{\partial V}{\partial q_2}+\frac{\partial D}{\partial \dot{q}_2} = Q_2, \tag{7.17}$$

we obtain the following linear differential equation for q_2:

$$q_2+4{\cdot}57t_1\,\dot{q}_2 = \theta_0, \tag{7.18}$$

where t_1 is the transit time (7.11). We integrate equation (7.18) with the initial value $q_2 = 0$ for $t = t_1$. The result is

$$\frac{q_2}{\theta_0} = 1-\exp\left\{-0{\cdot}218\left(\frac{t}{t_1}-1\right)\right\}. \tag{7.19}$$

The approximate values of θ/θ_0 derived from equation (7.1) for the first phase and equation (7.12) for the second phase are plotted versus x/l in Fig. 1.2 for three values of the non-dimensional time parameter t/t_1. The dotted lines show the exact values obtained by the classical Fourier series expansion as given in the treatise by Carslaw and Jaeger.‡ As can be seen, the agreement is excellent.

The simplicity of the approximate solution as compared with the classical treatment is primarily due to the fact that the Fourier series is not a very suitable representation of the temperature distribution and is badly convergent in the initial phase. The present solution avoids this difficulty by considering two phases that are analysed by entirely different procedures. These procedures may then be chosen to suit the particular features of each phase.

A variety of transient heat conduction problems solved by the method used in the present example have been treated in the following publications.

1. M. A. Biot, 'New methods in heat flow analysis with application to flight structures', *J. aeronaut. Sci.* **24**, 857–73 (1957).
2. M. Levinson, 'Thermal stresses in an idealized wing structure', *J. Aerospace Sci.* **28**, 899–901 (1961).
3. T. J. Lardner, 'Biot's variational principle in heat conduction', *AIAA J.* **1**, 196–206 (1963).
4. H. N. Chu, 'Application of Biot's variational method to convective heating of a slab', *J. Spacecraft Rockets*, **1**, 686–8 (1964).
5. A. F. Emery, 'Use of Biot's variational technique in heat conduction', *AIAA J.* **3**, 1525–6 (1965).
6. Y. C. Fung, *Foundations of solid mechanics*, pp. 408–9. Prentice-Hall, Englewood Cliffs, New Jersey (1965).

‡ H. S. Carslaw and J. C. Jaeger, *Conduction of heat in solids*, p. 83. Clarendon Press, Oxford (1947).

CHAPTER TWO

GENERAL THEORY OF LINEAR SYSTEMS

1. INTRODUCTION

In this chapter we shall discuss the general theory of thermal systems with properties independent of the temperature. This covers a large class of physical systems that may be described by generalized coordinates and are governed by the same type of linear Lagrangian equations. As a consequence, it is possible to develop a unified theory of such systems‡ as shown in Section 2. This formulation includes, in particular, frequently encountered boundary conditions, which may be incorporated as part of the dissipation function of the complete system. The linear Lagrangian equations are expressed in Section 3. They are derived from the two quadratic forms that define the thermal potential and the dissipation function. Special attention is given to the properties of these forms as regards positive-definiteness since they have an important bearing on the behaviour of the solutions of the differential equations.

In the absence of thermal forces, the time history of the system is represented by thermal relaxation modes, as shown in Section 4. These modes are characteristic solutions, each of them proportional to a decreasing exponential function of time. The orthogonality property of the relaxation modes is derived in Section 5 along with the associated normal coordinates. The response of a system to given thermal forces is expressed in closed form in terms of normal coordinates. In the discussion, special attention has been given to the important singular cases of multiple characteristic roots and zero roots. The zero roots are shown to correspond to steady-state heat flow.

When a system is subject to given driving thermal forces we may consider the steady state that would correspond to an instantaneous value of these thermal forces assumed to remain constant. When the thermal forces vary, we may use a continuous sequence of steady state flows as part of the solution. Such a solution, valid for infinitely slow variations, may be termed quasi-steady. We then add a correction

‡ This unified approach was outlined in two papers by the author: 'New methods in heat flow analysis with application to flight structures', *J. aeronaut. Sci.* **24**, 857–73 (1957); and 'Further developments of new methods in heat flow analysis', *J. Aerospace Sci.* **26**, 367–81 (1959).

defined by generalized coordinates as developed in Section 6. Normal coordinates may be conveniently used for this purpose. As pointed out, solutions of this type will be particularly suitable for cases of slowly varying temperatures, since they tend to approach the case of quasi-steady flow.

Section 7 deals with the problem of transient heat flow through a slab as an illustration of the properties of normal coordinates. The example provides an opportunity to discuss the physical significance of so called 'weak' solutions that may result from the use of generalized coordinates.

2. BOUNDARY DISSIPATION FUNCTION

Until now we have not referred explicitly to boundary conditions. Surface temperatures were considered to be given functions of time and location leading to the generalized thermal forces. They may be looked upon as driving forces applied to the system. In many cases the surface temperatures are not prescribed. Instead, certain heat transfer properties of the boundary are given.

We shall consider here a very special case of linear boundary condition. It will be assumed that the local rate of heat flow per unit area of the surface is proportional to the difference between the temperature θ of the surface and the temperature θ_a of the surrounding medium. We write

$$\dot{H}_n = K(\theta - \theta_a). \tag{2.1}$$

The rate of outward heat flow per unit area of the boundary is denoted by $\dot{H}_n$ and K is a surface heat transfer coefficient. The temperature θ_a is defined by the property that if $\theta = \theta_a$ there is no heat flow at the surface. θ_a may be referred to as the *adiabatic surface temperature.* Equation (2.1) simply states that the heat flow is proportional to the deviation from the adiabatic temperature.

The surface heat transfer law (2.1) constitutes an approximation whose validity depends greatly on the type of problem considered. For example, the surface heat transfer may be due to radiation with small deviations from the adiabatic surface temperatures. In such a case equation (2.1) often constitutes a justified linearization of the radiation law. On the other hand, the concept of local heat transfer coefficient K is not physically correct for describing the heat flow between a solid wall and a moving fluid.‡ Use of equation (2.1) in this case will therefore be

‡ See, for example, the author's paper 'Fundamentals of boundary-layer heat transfer with streamwise temperature variations', *J. Aerospace Sci.* **29**, 558–67 (1962).

limited to certain categories of problems and only after careful examination of the approximations involved.

The more general case of heat transfer in a moving fluid will be discussed in Chapters 6 and 7.

The point of interest in connection with the surface heat transfer law (2.1) lies in the possibility of incorporating it into the dissipation function. In order to show this let us write the Lagrangian equations (5.23) of Chapter 1 in the form

$$\frac{\partial V}{\partial q_i}+\frac{\partial D'}{\partial \dot{q}_i} = Q_i'. \tag{2.2}$$

The dissipation function of the solid, as given by equation (5.18) of Chapter 1 for the most general case of anisotropic conductivity, is

$$D' = \tfrac{1}{2}\iiint_\tau \sum^{jk} \lambda_{jk}\dot{H}_j\dot{H}_k\,d\tau. \tag{2.3}$$

The thermal force Q_i' is defined by the temperature at the boundary A of the volume τ. Following equation (5.22) of Chapter 1 we write, in vector form,

$$Q_i' = -\iint_A \theta\frac{\partial \mathbf{H}}{\partial q_i}.\,\mathbf{n}\,dA. \tag{2.4}$$

The value

$$\theta = \frac{1}{K}\dot{H}_n+\theta_a \tag{2.5}$$

derived from equation (2.1) is now substituted in expression (2.4). This yields

$$Q_i' = -\iint_A \frac{1}{K}\dot{H}_n\frac{\partial \mathbf{H}}{\partial q_i}.\,\mathbf{n}\,dA-\iint_A \theta_a\frac{\partial \mathbf{H}}{\partial q_i}.\,\mathbf{n}\,dA. \tag{2.6}$$

Let us examine the first term. Because of equation (4.2) of Chapter 1 we may write

$$\iint_A \frac{1}{K}\dot{H}_n\frac{\partial \mathbf{H}}{\partial q_i}.\,\mathbf{n}\,dA = \iint_A \frac{1}{K}\dot{H}_n\frac{\partial \dot{H}_n}{\partial \dot{q}_i}\,dA. \tag{2.7}$$

Hence by putting

$$D'' = \frac{1}{2}\iint_A \frac{1}{K}\dot{H}_n^2\,dA \tag{2.8}$$

and

$$Q_i = -\iint_A \theta_a\frac{\partial \mathbf{H}}{\partial q_i}.\,\mathbf{n}\,dA \tag{2.9}$$

we derive

$$Q_i' = -\frac{\partial D''}{\partial \dot{q}_i}+Q_i. \tag{2.10}$$

Expression D'' is analogous to D' and represents a dissipation function for the boundary heat transfer. The total dissipation function is

$$D = D'+D''. \tag{2.11}$$

With the values (2.3) and (2.8) this is written explicitly as

$$D = \frac{1}{2}\iiint\limits_{\tau} \sum^{jk} \lambda_{jk}\dot{H}_j\dot{H}_k\,d\tau+\frac{1}{2}\iint\limits_{A}\frac{1}{K}\dot{H}_n^2\,dA. \tag{2.12}$$

For isotropic thermal conductivity we obtain the simpler form

$$D = \frac{1}{2}\iiint\limits_{\tau}\frac{1}{k}\dot{\mathbf{H}}^2\,d\tau+\frac{1}{2}\iint\limits_{A}\frac{1}{K}\dot{H}_n^2\,dA. \tag{2.13}$$

Using expressions (2.12) or (2.13) for D, we write equation (2.2) in the form

$$\frac{\partial V}{\partial q_i}+\frac{\partial D}{\partial \dot{q}_i} = Q_i. \tag{2.14}$$

In this equation the total dissipation function D includes both the volume dissipation D' and the boundary dissipation D''. The thermal force (2.9) driving the system is now expressed in terms of the adiabatic temperature θ_a.

The particular type of surface heat transfer considered here may be visualized by noting that everything occurs as if the surface were covered by a thin layer of material of zero heat capacity which separates the solid from a surrounding medium. The local temperature of the surrounding medium is θ_a and the local heat transmitting property of the thin layer is characterized by a coefficient K, depending on the thickness and thermal conductivity of the layer.

We note that the foregoing results are valid in the most general case where $K(x,y,z,t)$ may be a function of the time t and of the coordinates x, y, z at the boundary.

3. LINEAR LAGRANGIAN EQUATIONS

The generalized coordinates q_i may be chosen in such a way that they are linearly related to the field **H**. We write

$$\mathbf{H} = \sum^{i} \mathbf{H}^{(i)}(x,y,z)q_i, \tag{3.1}$$

where $\mathbf{H}^{(i)}(x,y,z)$ are given field configurations independent of time. There may be an infinite number of generalized coordinates, in which case expression (3.1) is an infinite series of vector fields. The fields $\mathbf{H}^{(i)}$

may be chosen arbitrarily, provided they are capable of representing the physics of the problem with suitable accuracy. In the mathematical sense, they may be chosen to represent the field with any degree of accuracy in the vicinity of a set of N points P_k in any given domain. The number N of these points may be arbitrarily large. The $3N$ components of the field **H** at the points P_k are functions of $3N$ generalized coordinates q_k through the linear transformation (3.1). The only requirement is that the transformation be non-singular. Hence, from the viewpoint of the physicist, the generalized coordinate representation of the continuous field does not restrict its generality.

We may apply the linear mathematical formulation (3.1) to a system with properties independent of the temperature. The problem is then linear both physically and mathematically. Since the heat capacity, the thermal conductivity, and the surface heat transfer coefficient are all independent of the temperature, the thermal potential and the dissipation function are quadratic forms with constant coefficients. They are obtained as follows.

The temperature field θ is derived from the heat displacement field **H** by equation (2.2) of Chapter 1. We write

$$\theta = -\frac{1}{c}\operatorname{div}\mathbf{H}. \tag{3.2}$$

Introducing the representation (3.1), the temperature is

$$\theta = \sum^{i} \theta^{(i)}(x, y, z) q_i \tag{3.3}$$

with

$$\theta^{(i)} = -\frac{1}{c}\operatorname{div}\mathbf{H}^{(i)}. \tag{3.4}$$

The field θ is thus expressed by a superposition of scalar configuration fields $\theta^{(i)}$ with amplitude factors proportional to the generalized coordinates q_i. The thermal potential is

$$V = \tfrac{1}{2} \iiint_{\tau} c\theta^2 \, d\tau. \tag{3.5}$$

Substitution of the value (3.3) for θ yields the quadratic form

$$V = \tfrac{1}{2} \sum^{ij} a_{ij} q_i q_j. \tag{3.6}$$

The coefficients are

$$a_{ij} = \iiint_{\tau} c\theta^{(i)}\theta^{(j)} \, d\tau. \tag{3.7}$$

For isotropic thermal conductivity the dissipation function is given by equation (2.13) as

$$D = \frac{1}{2}\iiint_{\tau} \frac{1}{k}\dot{\mathbf{H}}^2\,d\tau + \frac{1}{2}\iint_{A} \frac{1}{K}\dot{H}_n^2\,dA. \tag{3.8}$$

This includes the dissipation in the volume τ of the solid and the dissipation at the boundary A due to a surface heat transfer coefficient K. By introducing expression (3.1) for $\mathbf{H}$, the dissipation function (3.8) becomes the quadratic form

$$D = \tfrac{1}{2}\sum^{ij} b_{ij}\dot{q}_i\dot{q}_j \tag{3.9}$$

and the coefficients are

$$b_{ij} = \iiint_{\tau} \frac{1}{k}\mathbf{H}^{(i)}.\mathbf{H}^{(j)}\,d\tau + \iint_{A} \frac{1}{K}H_n^{(i)}H_n^{(j)}\,dA. \tag{3.10}$$

For anisotropic thermal conductivity we use the value (2.12) for the dissipation function. The coefficients b_{ij} in this case are

$$b_{ij} = \frac{1}{2}\iiint_{\tau} \sum^{kl} \lambda_{kl} H_k^{(i)}H_l^{(j)}\,d\tau + \frac{1}{2}\iint_{A} \frac{1}{K}H_n^{(i)}H_n^{(j)}\,dA. \tag{3.11}$$

According to equations (2.9), the generalized thermal forces are

$$Q_i = -\iint_{A} \theta_a \mathbf{H}^{(i)}.\mathbf{n}\,dA. \tag{3.12}$$

With the values (3.6) and (3.9), the Lagrangian equations

$$\frac{\partial V}{\partial q_i} + \frac{\partial D}{\partial \dot{q}_i} = Q_i \tag{3.13}$$

become

$$\sum^{j} a_{ij}q_j + \sum^{j} b_{ij}\dot{q}_j = Q_i. \tag{3.14}$$

This is a system of linear differential equations with constant coefficients for the generalized coordinates.

By definition of the quadratic forms (3.6) and (3.9), the coefficients satisfy the symmetry relations

$$a_{ij} = a_{ji}, \qquad b_{ij} = b_{ji}. \tag{3.15}$$

Let us examine further some important restrictions imposed upon the mathematical nature of the quadratic forms. Consider the dissipation function

$$D = D' + D'', \tag{3.16}$$

where D' and D'' are the separate dissipation functions for the solid and tbe boundary respectively as given by equations (2.3) and (2.8). These expressions cannot be negative and cannot vanish unless the rate of heat flow $\dot{\mathbf{H}}$ vanishes everywhere in the solid and at the boundary.

This is a consequence of the fact that the expression $\sum^{ij} \lambda_{kj} \dot{H}_k \dot{H}_j$ is proportional to the local entropy production, which is positive-definite. Hence, when expressed as a function of the heat displacement, the dissipation function is positive-definite. In addition, let us assume the transformation (3.1) to be non-singular, meaning by this that the field **H** cannot vanish unless all the generalized coordinates also vanish. In this case the dissipation function expressed as a function of $\dot{q}_i$ is also *positive-definite.*

The same property does not apply to the thermal potential. According to its definition it is non-negative but it may be zero for values of the field **H** for which $\theta = 0$. According to equation (2.2) of Chapter 1 this will occur for components of the field **H** such that

$$\operatorname{div} \mathbf{H} = 0. \tag{3.17}$$

The thermal potential may therefore vanish for values of the generalized coordinates which are not zero. In the mathematical terminology it is said that the thermal potential may be *positive-semidefinite.*

These properties along with the symmetry relations (3.15) are fundamental in the analysis of the general behaviour of linear thermal systems, as developed in the following sections.

4. THERMAL RELAXATION MODES

When the thermal forces Q_i vanish, equations (3.14) become

$$\sum^{j} a_{ij} q_j + \sum^{j} b_{ij} \dot{q}_j = 0. \tag{4.1}$$

They are linear and homogeneous. Physically, this may represent a solid where the initial temperature is different, inside the body, from the surface temperature maintained constant. Since the origin of the temperature scale is arbitrary the value zero is chosen to represent the surface temperature.

In the case where the boundary properties are represented by a surface heat transfer coefficient, the solid and its boundary are considered as a single system. As shown in Section 2, the properties of the boundary are taken into account by adding a suitable term in the expression of the dissipation function. A vanishing thermal force in this case corresponds to a solid when the surrounding medium is maintained at constant temperature. This is expressed by putting equal to zero the adiabatic temperature θ_a.

Finally, the thermal force Q_i will also vanish if some portions of the boundary are impervious to heat flow. This can be seen from expression (2.9), since $(\partial\mathbf{H}/\partial q_i).\mathbf{n}$ will vanish at points where there is no heat flow across the boundary.

For physical reasons the solution of equations (4.1) must represent a temperature field that vanishes asymptotically as it tends toward equilibrium. Consider solutions of the type

$$q_i = C_i e^{pt}, \tag{4.2}$$

where C_i and p are constants. By substitution in equations (4.1) we derive

$$\sum^{j} (a_{ij}+pb_{ij})C_j = 0. \tag{4.3}$$

Equations (4.3) constitute a system of linear algebraic equations in the variables C_j with an unknown parameter p. Compatibility of these equations is expressed by putting their determinant equal to zero. We write

$$\det|a_{ij}+pb_{ij}| = 0. \tag{4.4}$$

If there are n variables this is an algebraic equation of degree n in p called the *characteristic equation*. A classical result of matrix theory states that all roots must be real. This is a consequence of the two following assumptions:

(1) the coefficients a_{ij} and b_{ij} are symmetric ($a_{ij} = a_{ji}$ and $b_{ij} = b_{ji}$);

(2) the quadratic form $D = \frac{1}{2}\sum^{ij} b_{ij}\dot{q}_i\dot{q}_j$ is positive-definite.

That the roots p must be real can be shown by assuming a complex value p and its conjugate p^*. Corresponding complex values C_j and C_j^* satisfy the equations

$$\sum^{j}(a_{kj}+pb_{kj})C_j = 0, \qquad \sum^{k}(a_{kj}+p^*b_{kj})C_k^* = 0, \tag{4.4 a}$$

Multiply the first equations by C_k^* and sum over all values of k. Similarly, multiply the second equations by C_j and sum over all values of j. The difference of the two results yields

$$(p-p^*)\sum^{kj} b_{kj}C_j C_k^* = 0. \tag{4.4 b}$$

If we put $C_j = \alpha_j + i\beta_j$ we may write

$$\sum^{kj} b_{kj}C_j C_k^* = \sum^{kj} b_{kj}(\alpha_k\alpha_j+\beta_k\beta_j). \tag{4.4 c}$$

Since $D = \frac{1}{2}\sum^{kj} b_{kj}\dot{q}_k\dot{q}_j$ is positive-definite expression (4.4 c) is also positive. According to equation (4.4 b) this implies $p = p^*$. Hence the values of p are real.

In addition, the roots must be non-positive. This can be seen by considering a root p and its corresponding real constants C_j satisfying equations (4.3). We multiply these equations by C_i and, after summation, we obtain

$$\sum^{ij} (a_{ij}+pb_{ij})C_i C_j = 0. \tag{4.5}$$

Hence
$$p = -\frac{V_c}{D_i}, \tag{4.6}$$
where
$$V_c = \sum^{ij} a_{ij} C_i C_j, \qquad D_c = \sum^{ij} b_{ij} C_i C_j. \tag{4.7}$$
Now D_c is positive-definite and V_c is positive-semidefinite. Therefore p is either zero or negative.

Consider a system with n generalized coordinates q_i. The characteristic equation (4.4) is then of degree n. Let us assume that the *roots are all distinct*. We may write the n values of these roots in the form
$$p = -\lambda_s, \tag{4.8}$$
where $s = 1, 2, \ldots, n$ and λ_s is either positive or zero. The general solution of the differential equation (4.1) is then
$$q_i = \sum^{s} C_i^{(s)} e^{-\lambda_s t}. \tag{4.9}$$
The n *characteristic solutions* $C_i^{(s)} \exp(-\lambda_s t)$ are the *thermal relaxation modes*. They represent thermal fields of fixed configuration each with an exponentially decaying amplitude. The coefficients λ_s are the *relaxation constants*.

For a given value of s the constants $C_i^{(s)}$ are determined only by their ratios. To bring this out we write
$$C_i^{(s)} = C^{(s)} \varphi_i^{(s)}. \tag{4.10}$$
Where $C^{(s)}$ is an arbitrary amplitude and $\varphi_i^{(s)}$ are fixed values satisfying a certain *normalizing condition*. It is convenient in the present case to choose as normalizing condition the following relation:
$$\sum^{ij} b_{ij} \varphi_i^{(s)} \varphi_j^{(s)} = 1. \tag{4.11}$$
This choice is motivated by the fact that D is positive-definite. The general solution (4.9) now takes the form
$$q_i = \sum^{s} C^{(s)} \varphi_i^{(s)} e^{-\lambda_s t} \tag{4.12}$$
as a sum of normalized relaxation modes $\varphi_i^{(s)} \exp(-\lambda_s t)$, each with an arbitrary amplitude $C^{(s)}$.

When the *characteristic roots* $p = -\lambda_s$ are all distinct, the existence of a general solution of the form (4.12) is a consequence of a classical theorem for differential equations with constant coefficients. When there are multiple characteristic roots this is not generally true,‡ and the general solution may contain terms of the type $t^k \exp(-\lambda_s t)$. However,

‡ See, for example, F. R. Moulton, *Differential equations*, chapter xv, Macmillan, New York (1930).

in the present case, which is limited to the particular case of differential equations (4.1), it is possible to show that the general solution (4.12) containing only exponential terms remains valid also if there are multiple roots. This is a consequence of the symmetry of the coefficients ($a_{ij} = a_{ji}$, $b_{ij} = b_{ji}$) and of the positive-definiteness of the dissipation function D. A more detailed discussion of this case is given in the next section in connection with normal coordinates.

Variational principle for relaxation modes

Consider the quadratic forms V_c and D_c defined by equations (4.7). They are functions of the variables C_i. We ask for stationary values of V_c, hence we put

$$\delta V_c = 0 \tag{4.13}$$

for variations δC_i satisfying the condition that D_c is constant. This is an extremum problem with the constraint

$$D_c = \text{const.}, \tag{4.14}$$

a standard problem of the variational calculus. Its solution is obtained by equating to zero the unconstrained variation

$$\delta(V_c + pD_c) = 0, \tag{4.15}$$

where p is a constant representing an undetermined Lagrangian multiplier. Equation (4.15) is written explicitly as

$$\sum^{ij} (a_{ij} + pb_{ij})C_j \delta C_i = 0. \tag{4.16}$$

It must be verified for arbitrary values of the variations δC_i. This leads to the relations

$$\sum^{j} (a_{ij} + pb_{ij})C_j = 0, \tag{4.17}$$

which are identical to equations (4.3). Hence the variational principle defined by equations (4.13) and (4.14) provides an equivalent derivation of the relaxation modes. The acceptable values of the Lagrangian multiplier p represent the roots of the characteristic equation (4.4).

5. ORTHOGONALITY AND NORMAL COORDINATES

Consider two relaxation modes:

$$q_i = C_i^{(s)} e^{-\lambda_s t}, \qquad q_i = C_i^{(r)} e^{-\lambda_r t}, \tag{5.1}$$

with different relaxation constants

$$\lambda_s \neq \lambda_r. \tag{5.2}$$

They satisfy the equations

$$\begin{aligned} \sum^{j} (a_{ij}\, C_j^{(s)} - \lambda_s b_{ij}\, C_j^{(s)}) &= 0, \\ \sum^{j} (a_{ij}\, C_j^{(r)} - \lambda_r b_{ij}\, C_j^{(r)}) &= 0. \end{aligned} \tag{5.3}$$

We multiply the first set of equations (5.3) by $C_i^{(r)}$ and sum the results. We obtain

$$\sum^{ij} (a_{ij}\, C_j^{(s)}\, C_i^{(r)} - \lambda_s b_{ij}\, C_j^{(s)}\, C_i^{(r)}) = 0. \tag{5.4}$$

Similarly, multiplying the second set of equations (5.3) by $C_i^{(s)}$, we obtain

$$\sum^{ij} (a_{ij}\, C_j^{(r)} C_i^{(s)} - \lambda_r b_{ij}\, C_j^{(r)} C_i^{(s)}) = 0. \tag{5.5}$$

We now subtract equation (5.5) from equation (5.4), taking into account the symmetry property of the coefficients ($a_{ij} = a_{ji}$, $b_{ij} = b_{ji}$). This yields

$$(\lambda_s - \lambda_r) b_{ij}\, C_i^{(r)} C_j^{(s)} = 0. \tag{5.6}$$

Since $\lambda_r \neq \lambda_s$, this result implies

$$\sum^{ij} b_{ij}\, C_i^{(r)} C_j^{(s)} = 0. \tag{5.7}$$

Because of equation (5.4) this result also implies

$$\sum^{ij} a_{ij}\, C_i^{(r)} C_j^{(s)} = 0. \tag{5.8}$$

Relations (5.7) and (5.8) are the well-known *orthogonality conditions* satisfied by characteristic solutions of systems defined by a pair of quadratic forms.

The orthogonality conditions may be expressed in terms of the normalized amplitudes $\varphi_i^{(s)}$ of the relaxation modes as defined by the normalizing condition (4.11). By substituting the values (4.10) into relations (5.7) and (5.8), the orthogonality conditions of the relaxation modes become

$$\sum^{ij} b_{ij}\, \varphi_i^{(r)} \varphi_j^{(s)} = 0, \qquad \sum^{ij} a_{ij}\, \varphi_i^{(r)} \varphi_j^{(s)} = 0. \tag{5.9}$$

In deriving the orthogonality property we have assumed $\lambda_r \neq \lambda_s$. However, the property may be extended to modes with the same relaxation constants. Consider, for example, k modes

$$\varphi_i^{(s)} \quad (s = m+1,\, m+2,\ldots,\, m+k)$$

appearing in the general solution (4.12), with the same relaxation constant λ. As can be shown, the choice of $\varphi_i^{(s)}$ to represent these modes is not unique and any system of independent base vectors may be used. It is readily verified that they may always be chosen so that they satisfy the first of relations (5.9). Because of equation (5.4) the second of relations (5.9) is also verified.

Let us introduce the scalar and vector fields,

$$\theta^{(s)} = \sum^{i} \theta^{(i)} \varphi_i^{(s)}, \qquad \mathbf{H}^{(s)} = \sum^{i} \mathbf{H}^{(i)} \varphi_i^{(s)}, \tag{5.10}$$

where $\theta^{(i)}$ and $\mathbf{H}^{(i)}$ are the fields in equations (3.3) and (3.1). With this definition the temperature and heat displacement fields of a relaxation mode are

$$\theta = C^{(s)} \theta^{(s)} e^{-\lambda_s t}, \qquad \mathbf{H} = C^{(s)} \mathbf{H}^{(s)} e^{-\lambda_s t}. \tag{5.11}$$

Hence $\theta^{(s)}$ and $\mathbf{H}^{(s)}$ may be considered as normalized fields defining the characteristic solutions.

Consider two temperature fields $\theta^{(r)}$ and $\theta^{(s)}$, corresponding to relaxation modes with different relaxation constants. We find that

$$\iiint_{\tau} c\theta^{(r)} \theta^{(s)} \, d\tau = 0. \tag{5.12}$$

This is readily verified by substituting expression (5.10) for $\theta^{(s)}$ and $\theta^{(r)}$, taking into account the value (3.7) for a_{ij} and the orthogonality condition (5.9).

Similarly, for the case of isotropic thermal conductivity, using the value (3.10) for b_{ij}, we derive

$$\iiint_{\tau} \frac{1}{k} \mathbf{H}^{(r)} . \mathbf{H}^{(s)} \, d\tau + \iint_{A} \frac{1}{K} H_n^{(r)} H_n^{(s)} \, dA = 0. \tag{5.13}$$

Equations (5.12) and (5.13) therefore express the orthogonality properties of the relaxation modes by means of the fields that characterize these modes.

For anisotropic thermal conductivity the orthogonality condition (5.13) is replaced by

$$\iiint_{\tau} \sum^{kj} \lambda_{kj} H_k^{(r)} H_j^{(s)} \, d\tau + \iint_{A} \frac{1}{K} H_n^{(r)} H_n^{(s)} \, dA = 0. \tag{5.14}$$

Normal coordinates

Consider a linear system described by n generalized coordinates q_i and the two quadratic forms

$$V = \tfrac{1}{2} \sum^{ij} a_{ij} q_i q_j, \qquad D = \tfrac{1}{2} \sum^{ij} b_{ij} \dot{q}_i \dot{q}_j. \tag{5.15}$$

The form D is positive-definite. A fundamental theorem of algebra‡

‡ See, for example, M. Bocher, *Introduction to higher algebra*, p. 170. Macmillan, New York (1924).

states that if D is positive-definite it is possible to find a non-singular transformation of the n variables q_i into n variables ξ_i in such a way that V and D become sums of squares of ξ_i. This transformation is written as

$$q_i = \sum^s \varphi_i^{(s)} \xi_s. \tag{5.16}$$

By a non-singular transformation is meant that the $n \times n$ determinant of the coefficients $\varphi_i^{(s)}$ does not vanish. Hence the values of q_i cannot be zero unless all the values of ξ_s are also zero. The theorem states that there exists a transformation (5.16) such that V and D become

$$V = \tfrac{1}{2} \sum^s \lambda_s \xi_s^2, \qquad D = \tfrac{1}{2} \sum^s \dot{\xi}_s^2. \tag{5.17}$$

Furthermore, the values of λ_s are the n roots of the characteristic equation

$$\det|a_{ij} - \lambda b_{ij}| = 0. \tag{5.18}$$

All the roots of this equation are real. Since it is assumed here that V is positive semidefinite the values of λ_s are non-negative. However, some of these values may be zero, hence V may contain less than n squares.

The theorem does not require that the values λ_s be distinct. There may be any number of equal values corresponding to multiple roots of the characteristic equation with any degree of multiplicity. The variables ξ_s define the *normal coordinates*. With these variables the Lagrangian equations in the absence of thermal forces are

$$\frac{\partial V}{\partial \xi_s} + \frac{\partial D}{\partial \dot{\xi}_s} = 0. \tag{5.19}$$

Introducing the values (5.17) yields n uncoupled differential equations

$$\lambda_s \xi_s + \dot{\xi}_s = 0. \tag{5.20}$$

The solution of these equations is

$$\xi_s = C^{(s)} e^{-\lambda_s t}. \tag{5.21}$$

They represent the thermal relaxation modes already introduced in Section 4. By substituting the values (5.21) of ξ_s into the transformation (5.16) we obtain

$$q_i = \sum^s C^{(s)} \varphi_i^{(s)} e^{-\lambda_s t}, \tag{5.22}$$

which coincides with expression (4.12).

This result also establishes the property, already stated in Section 4, that the general solution of the system of differential equations (4.1) is a sum of real exponentials and that this remains the case for any degree of multiplicity of the characteristic roots. We note that for vanishing roots ($\lambda_s = 0$) the corresponding exponential terms degenerate into constant values $C^{(s)} \varphi_i^{(s)}$.

In case there are multiple characteristic roots there is an indeterminacy in the choice of the coefficients $\varphi_i^{(s)}$. To illustrate this, consider the case where three values of the characteristic roots are equal, say

$$\lambda_1 = \lambda_2 = \lambda_3. \tag{5.23}$$

Let us restrict expressions (5.17) to the terms containing the corresponding normal coordinates ξ_1, ξ_2, ξ_3. We write

$$V = \tfrac{1}{2}\lambda_1(\xi_1^2+\xi_2^2+\xi_3^2), \qquad D = \tfrac{1}{2}(\dot{\xi}_1^2+\dot{\xi}_2^2+\dot{\xi}_3^2). \tag{5.24}$$

We may consider ξ_1, ξ_2, ξ_3, as rectangular coordinates in three-dimensional space. A rotation of the coordinate system transforms ξ_1, ξ_2, ξ_3, into new variables ξ_1', ξ_2', ξ_3', such that

$$\xi_1^2+\xi_2^2+\xi_3^2 = \xi_1'^2+\xi_2'^2+\xi_3'^2. \tag{5.25}$$

Hence with these new variables the quadratic forms (5.17) remain unchanged. The same indeterminacy occurs whenever there is a multiple root of any degree of multiplicity.

By substituting the transformation (5.16) into the quadratic forms (5.15) they are reduced to expressions (5.17). Since the cross-products must vanish this requires that the following relations be verified:

$$\begin{aligned} \sum^{ij} a_{ij}\varphi_i^{(r)}\varphi_j^{(s)} &= 0, \\ \sum^{ij} b_{ij}\varphi_i^{(r)}\varphi_j^{(s)} &= 0, \end{aligned} \qquad (r \neq s). \tag{5.26}$$

They are the same as the orthogonality conditions (5.9) derived for the case where $\lambda_r \neq \lambda_s$. Here equations (5.26) are derived in a different way as a consequence of the transformation (5.16). This shows that the orthogonality conditions (5.26) are valid for multiple roots, hence do not require that the values of λ_r and λ_s be distinct.

The value (5.17) of D also requires that the coefficients $\varphi_i^{(s)}$ in the transformation (5.16), verify the relations

$$\sum^{ij} b_{ij}\varphi_i^{(s)}\varphi_j^{(s)} = 1. \tag{5.27}$$

This coincides with the normalizing condition (4.11) for the relaxation modes.

Consider now the case of a system driven by thermal forces Q_i. The generalized coordinates obey the system of differential equations

$$\sum^{j} a_{ij}q_j + \sum^{j} b_{ij}\dot{q}_j = Q_i. \tag{5.28}$$

These equations may be expressed in terms of normal coordinates ξ_s

by the transformation (5.16). The Lagrangian equations become

$$\frac{\partial V}{\partial \xi_s}+\frac{\partial D}{\partial \dot{\xi}_s} = \Xi_s, \tag{5.29}$$

where V and D are given by equations (5.17) and Ξ_s is the generalized force conjugate to the normal coordinate ξ_s. The value of Ξ_s is found by applying equation (4.7) of Chapter 1. According to this equation we may write the identity

$$\sum^{i} Q_i \delta q_i = \sum^{s} \Xi_s \delta\xi_s. \tag{5.30}$$

By the transformation (5.16) the variations δq_i and $\delta\xi_s$ are related as follows:

$$\delta q_i = \sum^{s} \varphi_i^{(s)} \delta\xi_s. \tag{5.31}$$

By substituting these values into equation (5.30) and identifying the coefficients of $\delta\xi_s$ on both sides of the equation, we derive

$$\Xi_s = \sum^{i} \varphi_i^{(s)} Q_i. \tag{5.32}$$

This determines the normal forces in terms of the original values Q_i.

The differential equations (5.29) for the normal coordinates are written explicitly as

$$\lambda_s \xi_s+\dot{\xi}_s = \Xi_s. \tag{5.33}$$

When Q_i are given functions of time, the values $\Xi_s(t)$ are also known functions of time by relations (5.32). Let us assume that the system is initially undisturbed and that application of the forces Ξ_s is initiated at $t = 0$. Hence

$$\xi_s = \Xi_s = 0 \quad \text{for } t < 0. \tag{5.34}$$

In this case the solutions of equations (5.33) are completely determined and are given by the well-known expressions

$$\xi_s = e^{-\lambda_s t} \int\limits_0^t e^{\lambda_s t'} \Xi_s(t')\, dt'. \tag{5.35}$$

If thermal forces with constant values Ξ_s are suddenly applied at $t = 0$ the values (5.35) of ξ_s become

$$\xi_s = (1-e^{-\lambda_s t})\frac{\Xi_s}{\lambda_s}. \tag{5.36}$$

As already pointed out, a certain number of relaxation constants λ_s are zero. They correspond to components of the heat displacement field $\mathbf{H}$ for which $\operatorname{div}\mathbf{H} = 0$. The corresponding normal coordinates obey the differential equations

$$\dot{\xi}_s = \Xi_s. \tag{5.37}$$

When constant thermal forces Ξ_s are applied at $t = 0$ equations

(5.37) yield the solutions

$$\xi_s = \Xi_s t. \tag{5.38}$$

These normal coordinates represent the solution corresponding to a *steady-state flow* that is reached asymptotically for large values of the time t after all transients have died out. In steady-state flow in accordance with equations (5.38) the heat displacement **H** is a linear function of time.

6. QUASI-STEADY FLOW

Consider a system subject to thermal forces that vary very slowly. When this variation becomes infinitely slow the thermal field tends in the limit to a continuous sequence of steady states. This means that at any instant the temperature and heat flow are the same as if the thermal forces were constant and equal to their value at that particular instant. We shall refer to a solution of this type as a quasi-steady solution.

This concept suggests a way of solving transient heat-flow problems by considering the quasi-steady solution as a first approximation to which corrections are added to account for the transient behaviour. Solutions of this type should be more rapidly convergent, particularly in problems with slowly varying temperatures. This will be illustrated in the example treated at the end of the next section.

We denote by θ^* the temperature field of the quasi-steady solution. We shall assume that it has been evaluated independently and is a known function of time and location:

$$\theta^* = \theta^*(x, y, z, t). \tag{6.1}$$

It may be evaluated by any of the known methods to solve steady-state heat flow problems including, of course, variational methods. The actual temperature field of the system is

$$\theta = \theta^+ + \theta^*, \tag{6.2}$$

where θ^+ is the corrective term due to the transient behaviour. Let us assume isotropic thermal conductivity. Since θ^* is a quasi-steady temperature field it satisfies the equation

$$\operatorname{div}(k \operatorname{grad} \theta^*) = 0. \tag{6.3}$$

On the other hand, the actual temperature θ satisfies the heat-diffusion equation (2.10a) of Chapter 1, namely,

$$\operatorname{div}(k \operatorname{grad} \theta) = c\frac{\partial \theta}{\partial t}. \tag{6.4}$$

By substituting the value (6.2) for θ, taking into account equation (6.3), we derive

$$\operatorname{div}(k \operatorname{grad} \theta^+) = c\frac{\partial \theta^+}{\partial t} + c\frac{\partial \theta^*}{\partial t}. \tag{6.5}$$

Comparison with equation (6.19 a) of Chapter 1 shows that θ^+ represents the temperature in an analogue model with distributed heat sources of magnitude

$$w = -c\frac{\partial \theta^*}{\partial t}. \tag{6.6}$$

In addition, $\theta = \theta^*$ at the boundary, as follows from the definition of the steady-state solution. Hence in the analogue model with sources the boundary condition is

$$\theta^+ = 0. \tag{6.7}$$

The Lagrangian equations for this analogue model have been derived in Section 6 of Chapter 1. In the present case they are formulated as follows.

We choose a field $\mathbf{H}^*$ satisfying the equation

$$\operatorname{div} \mathbf{H}^* = \int_0^t w\,dt = -c\theta^*. \tag{6.8}$$

The field $\mathbf{H}^*$ is not determined uniquely by θ^*. The unknown field $\mathbf{H}^+$ representing the correction due to the transient behaviour is written

$$\mathbf{H}^+ = \mathbf{H}^+(q_1, q_2, \ldots, q_n, x, y, z, t) \tag{6.9}$$

as a function of the unknown generalized coordinates q_i. The temperature correction is

$$\theta^+ = -\frac{1}{c} \operatorname{div} \mathbf{H}^+. \tag{6.10}$$

According to equation (6.19) of Chapter 1, we may now write

$$\frac{\partial V^+}{\partial q_i} + \frac{\partial D^+}{\partial \dot{q}_i} = Q_i^+, \tag{6.11}$$

where

$$V^+ = \tfrac{1}{2} \iiint_\tau c(\theta^+)^2\,d\tau,$$

$$D^+ = \frac{1}{2} \iiint_\tau \frac{1}{k} (\dot{\mathbf{H}}^+)^2\,d\tau, \tag{6.12}$$

$$Q_i^+ = -\iiint_\tau \frac{1}{k} \frac{\partial \mathbf{H}^+}{\partial q_i} . \dot{\mathbf{H}}^*\,d\tau.$$

The value of Q_i vanishes in the analogue model because θ^+ is zero at the boundary. Relations (6.11) are the Lagrangian equations for the generalized coordinates which determine the correction field θ^+.

7. ILLUSTRATIVE EXAMPLE—WEAK SOLUTIONS

The diffusion of heat into a slab of thickness l with one face thermally insulated was treated in Section 7 of Chapter 1 as a one-dimensional problem by using the concept of penetration depth. In order to illustrate the procedures developed in the present chapter, we shall consider the heating of a slab, assuming both faces to be pervious to heat and using an entirely different procedure based on normal coordinates.

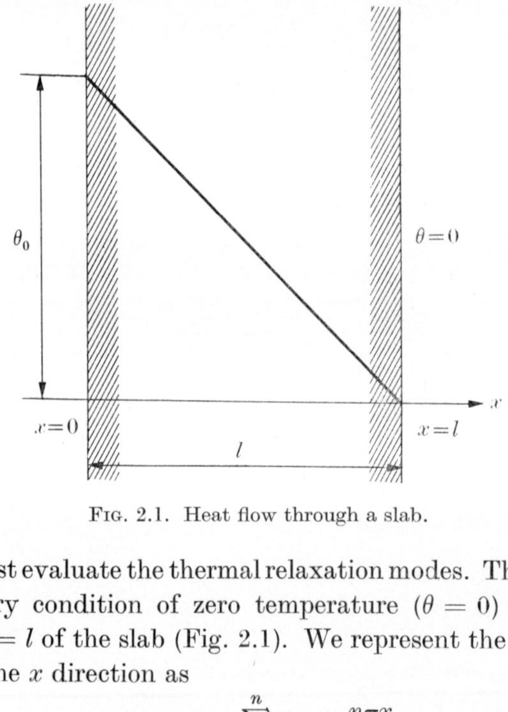

FIG. 2.1. Heat flow through a slab.

We shall first evaluate the thermal relaxation modes. They correspond to a boundary condition of zero temperature ($\theta = 0$) at both faces $x = 0$ and $x = l$ of the slab (Fig. 2.1). We represent the heat displacement along the x direction as

$$H = q_0 + \sum^{n} q_n \cos\frac{n\pi x}{l} \tag{7.1}$$

with n = positive integer. The corresponding temperature field $\theta = -(1/c)\, dH/dx$ is

$$\theta = \frac{\pi}{cl}\sum^{n} nq_n \sin\frac{n\pi x}{l}. \tag{7.2}$$

It satisfies the boundary condition $\theta = 0$ at $x = 0$ and $x = l$. It is sufficient in the present case to evaluate the thermal potential and the dissipation function for a volume equal to a cylinder of unit cross-section perpendicular to the slab. We write

$$V = \tfrac{1}{2}c\int\limits_0^l \theta^2\,dx, \qquad D = \frac{1}{2k}\int\limits_0^l \dot{H}^2\,dx. \tag{7.3}$$

Because of orthogonality properties of the trigonometric terms in expressions (7.1) and (7.2) the values (7.3) are reduced to sums of squares. We find

$$V = \frac{1}{4}\frac{\pi^2}{cl}\sum^n n^2 q_n^2, \qquad D = \frac{l}{2k}\dot{q}_0^2 + \frac{1}{4}\frac{l}{k}\sum^n \dot{q}_n^2. \tag{7.4}$$

In the absence of thermal forces the generalized coordinates q_0, q_n obey the Lagrangian equations

$$\frac{\partial V}{\partial q_n} + \frac{\partial D}{\partial \dot{q}_n} = 0, \qquad \frac{\partial D}{\partial \dot{q}_0} = 0. \tag{7.5}$$

By substituting the values (7.4) we derive

$$\lambda_n q_n + \dot{q}_n = 0, \qquad \dot{q}_0 = 0, \tag{7.6}$$

with

$$\lambda_n = \frac{\pi^2 k n^2}{cl^2}. \tag{7.7}$$

The solutions of equations (7.6) are

$$q_n = C_n e^{-\lambda_n t}, \qquad q_0 = \text{const.}, \tag{7.8}$$

with arbitrary constants C_n. These solutions represent the relaxation modes of the slab. We note the existence of a mode q_0 of constant value which corresponds to the case of a vanishing relaxation constant ($\lambda_r = 0$).

These results show that q_0 and q_n are proportional to the normal coordinates ξ_s discussed in Section 5. The fact that they do not satisfy the same normalizing condition as ξ_r is immaterial in the applications. Hence we may also refer to q_0 and q_n as normal coordinates.

As an illustration of the general procedure developed in Section 5, we may use these normal coordinates to derive the transient heat diffusion in the slab when the face at $x = 0$ is suddenly brought to a temperature $\theta = \theta_0$ at the time $t = 0$. We assume that the opposite face at $x = l$ is maintained at zero temperature.‡ Thermal forces Q_0 and Q_n must be added to the Lagrangian equations (7.5). They become

$$\frac{\partial V}{\partial q_n} + \frac{\partial D}{\partial \dot{q}_n} = Q_n, \qquad \frac{\partial D}{\partial \dot{q}_0} = Q_0. \tag{7.9}$$

‡ The analysis presented here follows the treatment of the same problem in the author's paper, 'Thermoelasticity and irreversible thermodynamics', *J. appl. Phys.* **27**, 240–53 (1956).

In order to obtain the thermal forces we must evaluate

$$\theta_0 \delta H = Q_0 \delta q_0 + \sum^n Q_n \delta q_n, \tag{7.10}$$

where δH is the variation of H at $x = 0$. According to equation (7.1) this variation is

$$\delta H = \delta q_0 + \sum^n \delta q_n. \tag{7.11}$$

By substituting this value into equation (7.10) and equating the coefficients of δq_0 and δq_n on both sides we derive

$$Q_0 = Q_n = \theta_0. \tag{7.12}$$

With these values, and expressions (7.4) for V and D, equations (7.9) are written as

$$\lambda_n q_n + \dot{q}_n = \frac{2k}{l}\theta_0, \qquad \dot{q}_0 = \frac{k}{l}\theta_0. \tag{7.13}$$

The solutions of these equations, with initial conditions $q_0 = q_n = 0$ at $t = 0$, are

$$q_n = \frac{2k\theta_0}{l\lambda_n}(1-e^{-\lambda_n t}), \qquad q_0 = \frac{k}{l}\theta_0 t. \tag{7.14}$$

Introduction of these values into expressions (7.1) and (7.2) yield for H and θ the time-dependent Fourier series:

$$\begin{aligned} H &= \frac{k}{l}\theta_0 t + \frac{2k\theta_0}{l}\sum^n \frac{1}{\lambda_n}(1-e^{-\lambda_n t})\cos\frac{n\pi x}{l}, \\ \theta &= \frac{2\theta_0}{\pi}\sum^n \frac{1}{n}(1-e^{-\lambda_n t})\sin\frac{n\pi x}{l}. \end{aligned} \tag{7.15}$$

As time increases these values tend toward the steady-state solution. For $t = \infty$, the limiting values of $\dot{H} = \partial H/\partial t$ and θ are

$$\dot{H} = \frac{k\theta_0}{l}, \qquad \theta = \frac{2\theta_0}{\pi}\sum^n \frac{1}{n}\sin\frac{n\pi x}{l} = \theta_0\left(1-\frac{x}{l}\right). \tag{7.16}$$

This limiting value of θ obtained here is the well-known Fourier series expansion for the linear function $\theta_0(1-x/l)$. This verifies an obvious result since this linear function represents the steady-state temperature distribution for $\theta = \theta_0$ at $x = 0$, and $\theta = 0$ at $x = l$. The limiting value $\dot{H} = k\theta_0/l$ is also the steady-state rate of heat flow across the wall. This illustrates the physical significance of the coordinate

$$q_0 = \frac{k}{l}\theta_0 t \tag{7.17}$$

as the linearly increasing heat displacement through the slab in the steady state.

Weak solutions

Attention is called to an important feature of the value (7.15) of θ. The Fourier series representation is such that $\theta = 0$ for $x = 0$. Hence the boundary condition $\theta = \theta_0$ is never satisfied at that point. However, for a point x infinitely close to the origin, the value jumps to $\theta = \theta_0$. Hence if we exclude the point $x = 0$, the boundary condition is satisfied. We are dealing here with a well-known case of non-uniform convergence of a Fourier series. In mathematical terminology a solution that breaks down at a finite number of points is said to be valid 'almost everywhere'. Such a solution is sometimes called a *weak solution*. The same terminology also designates cases where the solution breaks down on an infinite set of points provided the set has zero measure. Such weak solutions are typical of problems solved by using normal coordinates. However, this does not invalidate their physical significance, since the points at which they break down may be ignored because such a breakdown is physically spurious. This illustrates an important source of purely mathematical difficulty introduced into physical problems by the concept of continuum. Actually, from the standpoint of the physical scientist, it is a concept that contains many artificial features, and the difficulties will usually vanish by considering the purely physical aspect of the problem.

A solution that satisfies the boundary condition is obtained by writing the value (7.15) of θ in the form

$$\theta = \theta_0\left(1-\frac{x}{l}\right)-\frac{2\theta_0}{\pi}\sum^{n}\frac{1}{n}e^{-\lambda_n t}\sin\frac{n\pi x}{l}. \tag{7.18}$$

This solution could have been obtained directly by applying the procedure outlined in Section 6, starting from the steady-state solution and adding a correction represented by normal coordinates. This form of the solution converges very quickly for large values of t. As already pointed out, this type of solution combining a steady-state field and a transient correction will generally be preferable for slowly varying temperatures because of rapid convergence.

Another example of a solution of this type is given by expression (7.12) of Chapter 1, which is used to represent the second phase of the heating of a slab thermally insulated at $x = l$. The steady-state part of the solution in this case is a constant temperature θ_0 across the thickness.

CHAPTER THREE

OPERATIONAL FORMULATION

1. INTRODUCTION

THE operational formalism introduced by Heaviside provides a powerful analytical tool for the analysis of linear thermal systems. It is applicable to the large category of problems where the physical properties are independent of the temperature and the time. The operational approach provides a compact formulation that embodies several distinct but closely related mathematical theories. In particular, as shown by the author in the broader context of irreversible thermodynamics, it is possible to derive variational principles in operational form.

In Sections 2 and 3 we first develop the concepts of thermal admittance and impedance for harmonic dependence on the time. This leads to the concept of internal and external coordinates of thermal systems, and to general expressions for the thermal admittances and impedances which are partial fraction expansions.

Application of Fourier and Laplace transform methods to the evaluation of thermal transients is discussed in Section 4. How these methods lead to the basic operational formalism is outlined in Section 5. Particular attention is called to the use of 'generalized functions' in order to provide simplicity and generality for operational rules.

With these results it is possible to derive variational principles in operational form. As shown in Section 6, the compact formulation of these principles embodies distinct methods, corresponding to three different interpretations; one is an operational interpretation, the second an algebraic one, and the third corresponds to a convolution formulation.

As a corollary in Section 7, a variational principle is developed for interconnected systems. By this principle it is possible to formulate the equations of a composite system when the thermal impedances of the components are known. The result constitutes the analogue of a general theorem of mechanics whereby internal forces may be eliminated by a method of virtual work. A complementary form of this principle is also developed. A variety of 'finite element' methods are readily derived from these principles.

Section 8 provides an example of operational treatment of a one-dimensional problem in heat conduction. It also shows how a system of infinite extent leads to a thermal admittance expressed by means of a continuous spectrum of relaxation constants.

2. THERMAL ADMITTANCE

We consider a system subject to thermal forces that are harmonic functions of time. The generalized forces Q_i may be written $Q_i \exp(i\omega t)$, where Q_i now represents a complex amplitude. Similarly, the generalized coordinates q_i that represent the steady-state periodic response of the system must be replaced by $q_i \exp(i\omega t)$ with a complex amplitude q_i. Applying the general equations (3.14) of Chapter 2 for a linear system, the response q_i is found by solving the algebraic equations

$$\sum^{j} (a_{ij}+pb_{ij})q_j = Q_i, \tag{2.1}$$

where $p = i\omega$.

The physical significance of such a periodic solution is brought out by assuming that the harmonic forces are initiated at the instant $t = 0$ and are applied to a quiescent system, i.e. such that $q_i = 0$ for $t < 0$. The response is composed of a periodic solution and a transient. This transient is represented by expressions (4.12) of Chapter 2, which, in general, contain decreasing exponential and constant terms. Hence, after sufficient time, the steady-state response is represented by the harmonic solution superposed on constant values. The presence of such constant terms in the steady-state response is due to the fact that the transient term may produce a residual heat displacement that tends to a finite value as time increases. Mathematically, this is due to the semidefinite nature of the thermal potential and the existence of zero values of the relaxation constants.

Because the dissipation function is positive-definite the determinant of the coefficients b_{ij} is different from zero. Hence the determinant of equations (2.1) is also different from zero. We may, therefore, solve the equations for q_i. We write

$$q_i = \sum^{j} A_{ij} Q_j. \tag{2.2}$$

The coefficients A_{ij} constitute a complex *thermal admittance matrix* of the system. The relations $a_{ij} = a_{ji}$ and $b_{ij} = b_{ji}$ imply the property

$$A_{ij} = A_{ji}. \tag{2.3}$$

Hence the admittance matrix is symmetric.

General form of the admittance

The matrix elements are meromorphic functions of p. It is possible to derive a general expression for these functions by means of partial fractions.‡ This is accomplished by introducing the normal coordinates ξ_s. According to equations (5.16) of Chapter 2, the transformation from the normal coordinates to the coordinates q_i is written

$$q_i = \sum^{s} \varphi_i^{(s)} \xi_s. \tag{2.4}$$

The normal forces Ξ_s conjugate to these normal coordinates are given by equations (5.32) of Chapter 2. They are

$$\Xi_s = \sum^{i} \varphi_i^{(s)} Q_i. \tag{2.5}$$

The normal forces and normal coordinates are related by equations (5.33) of Chapter 2. For harmonic functions of the time these equations are written

$$\lambda_s \xi_s + p\xi_s = \Xi_s, \tag{2.6}$$

or

$$\xi_s = \frac{\Xi_s}{\lambda_s + p}. \tag{2.7}$$

By combining equations (2.4), (2.5), and (2.7), we obtain

$$q_i = \sum^{j} \sum^{s} \frac{\varphi_i^{(s)} \varphi_j^{(s)}}{\lambda_s + p} Q_j. \tag{2.8}$$

Comparing this result with equation (2.2) we see that the thermal admittance is a matrix of elements

$$A_{ij} = \sum^{s} \frac{\varphi_i^{(s)} \varphi_j^{(s)}}{\lambda_s + p}. \tag{2.9}$$

Hence we have obtained for A_{ij} a development in partial fractions. This value of A_{ij} may be expressed in a different way, which brings out the existence of multiple roots λ_s of the characteristic equation. Let us designate by λ_s a root of multiplicity k. In that case λ_s is repeated k times in the summation of equation (2.9). To this value of λ_s are attached k values $\varphi_i^{(s+s')}$, where s' takes values from 1 to k. We put

$$C_{ij}^{(s)} = \sum^{s'} \varphi_i^{(s+s')} \varphi_j^{(s+s')} \tag{2.10}$$

‡ The derivation follows the method used in the more general context of irreversible thermodynamics in the author's paper 'Theory of stress–strain relations in anisotropic viscoelasticity and relaxation phenomena', *J. appl. Phys.* **25**, 1385–91 (1954).

and write expression (2.9) in the form

$$A_{ij} = \sum^{s} \frac{C_{ij}^{(s)}}{\lambda_s + p}. \tag{2.11}$$

This time the summation $\sum^{s}$ is restricted to distinct values of λ_s.

An interesting property of the coefficients $C_{ij}^{(s)}$ is the non-negative character of the quadratic form associated with them. This property is brought out by considering the identity

$$\sum^{s'} \Big(\sum^{i} \varphi_i^{(s+s')} z_i\Big)^2 = \sum^{ij} C_{ij}^{(s)} z_i z_j, \tag{2.12}$$

which shows that the quadratic form on the right side cannot be negative.

Internal coordinates

In certain types of problems we may be interested only in the behaviour of a small number of coordinates in response to the conjugate forces. For example, thermal forces may be applied to a certain number of areas of the boundary. Consider k such areas denoted by $S_1, S_2,..., S_k$ and assume that temperatures $\theta_1, \theta_2,..., \theta_k$ are applied to each area. The temperatures are also assumed to be of constant amplitude over each area. They are complex quantities representing harmonic functions of time. The system may include a surface heat-transfer coefficient so that the applied thermal forces may be expressed in terms of adiabatic temperatures as defined in Section 2 of Chapter 2.

The generalized coordinates are separated into two groups. The average normal heat displacements $H_1, H_2,..., H_k$ distributed uniformly across each area $S_1, S_2,..., S_k$ may be chosen as generalized coordinates $q_1, q_2,..., q_k$. This first group will be called the *external coordinates*. If the displacements $H_1, H_2,..., H_k$ are chosen, positive inward the corresponding thermal forces are

$$Q_1 = S_1 \theta_1, \quad Q_2 = S_2 \theta_2, \quad ..., \quad Q_k = S_k \theta_k. \tag{2.13}$$

The complete system is represented by n generalized coordinates, $n > k$. There are therefore $n-k$ additional coordinates that describe the behaviour of the system. They are chosen in such a way that the corresponding heat displacement at the areas $S_1, S_2,..., S_k$ is either zero or such that its average normal component is zero for each area. We shall refer to this second group as the *internal coordinates*. The $n-k$ thermal forces conjugate to these internal coordinates are zero.

When the thermal forces $Q_1, Q_2,..., Q_k$ are applied to the areas $S_1, S_2,..., S_k$ the external coordinates may be expressed in terms of these thermal forces as

$$q_i = \sum_{j=1}^{k} A_{ij} Q_j \quad (i = 1, 2,..., k). \tag{2.14}$$

The internal coordinates do not appear in this formulation. The system behaves as a 'black box' and the internal structure of the system is represented only through the mathematical properties of the admittance elements A_{ij} as functions of p.

Reciprocity property

Consider two external areas S_1 and S_2 to which we apply respectively temperatures θ_1 and θ_2. The average heat displacements across the surfaces S_1 and S_2 are denoted by $q_1 = H_1$, $q_2 = H_2$ and chosen positive when directed inward. In this case the corresponding thermal forces are $Q_1 = S_1 \theta_1$, $Q_2 = S_2 \theta_2$. The heat displacement across S_2 due to the application of θ_1 at S_1 is

$$H_2 = A_{21} S_1 \theta_1. \tag{2.15}$$

Similarly, the heat displacement at S_1 due to the application of θ_2 at S_2 is

$$H_1 = A_{21} S_1 \theta_1. \tag{2.16}$$

The admittance elements satisfy the symmetry relation

$$A_{12} = A_{21}. \tag{2.17}$$

Hence, if $\theta_1 = \theta_2$, we derive

$$H_2 S_2 = H_1 S_1. \tag{2.18}$$

This result constitutes the reciprocity property.

3. THERMAL IMPEDANCE

Equations (2.14) for the k external coordinates $q_1, q_2,..., q_k$ in terms of the conjugate thermal forces $Q_1, Q_2,..., Q_k$ are, of course, quite general and apply whenever the remaining thermal forces are put equal to zero:

$$Q_{k+1} = Q_{k+2} = ... = Q_n = 0. \tag{3.1}$$

The admittance $[A_{ij}]$ of equations (2.14) is a $k \times k$ matrix. By solving these equations for Q_j we obtain

$$Q_i = \sum_{j=1}^{k} Z_{ij} q_j \quad (i = 1, 2,..., k). \tag{3.2}$$

The matrix of complex coefficients Z_{ij} will be called the *thermal impedance matrix*. Because $A_{ij} = A_{ji}$ we also have

$$Z_{ij} = Z_{ji}. \tag{3.3}$$

It is of interest to find an expression for Z_{ij} in the form of an expansion in partial fractions analogous to equation (2.11).

In order to derive these expressions‡ we consider the sub-system described by the coordinates $q_{k+1}, q_{k+2},..., q_n$. In describing the system we may replace these $n-k$ coordinates by the normal coordinates $\xi_{k+1}, \xi_{k+2},..., \xi_n$ of the sub-system. With these new coordinates the thermal potential and dissipation function of the total system are written

$$\begin{aligned} V &= \tfrac{1}{2}\sum^{ij} a_{ij} q_i q_j + \sum^{is} a'_{is} q_i \xi_s + \tfrac{1}{2}\sum^{s} r_s \xi_s^2, \\ D &= \tfrac{1}{2}\sum^{ij} b_{ij} \dot{q}_i \dot{q}_j + \sum^{is} b'_{is} \dot{q}_i \dot{\xi}_s + \tfrac{1}{2}\sum^{s} \dot{\xi}_s^2. \end{aligned} \tag{3.4}$$

In these summations, i and j vary from 1 to k and s varies from $k+1$ to n. With these new variables the Lagrangian equations separate into two groups, namely,

$$\begin{aligned} \frac{\partial V}{\partial q_i} + \frac{\partial D}{\partial \dot{q}_i} &= Q_i \quad (i = 1, 2,..., k), \\ \frac{\partial V}{\partial \xi_s} + \frac{\partial D}{\partial \dot{\xi}_s} &= \Xi_s \quad (s = k+1,..., n). \end{aligned} \tag{3.5}$$

The normal forces Ξ_s are given by equations (5.32) of Chapter 2. Hence,

$$\Xi_s = \sum^{m} \varphi_m^{(s)} Q_m, \tag{3.6}$$

where both s and m vary from $k+1$ to n. According to relations (3.1) the values Q_m are zero. This implies

$$\Xi_s = 0. \tag{3.7}$$

For harmonic time-dependence equations (3.5) are written explicitly as

$$\sum^{j} B_{ij} q_j + \sum^{s} B'_{is} \xi_s = Q_i, \qquad \sum^{i} B'_{is} q_i + (p+r_s)\xi_s = 0, \tag{3.8}$$

where

$$B_{ij} = a_{ij} + p b_{ij}, \qquad B'_{is} = a'_{is} + p b'_{is}. \tag{3.9}$$

We may solve for ξ_s the second group of equations (3.8); hence

$$\xi_s = -\sum^{i} \frac{B'_{is}}{p+r_s} q_i. \tag{3.10}$$

Substitution of this value into the first group of equations (3.8) yields

$$Q_i = \sum^{j} Z_{ij} q_j \tag{3.11}$$

with

$$Z_{ij} = B_{ij} - \sum^{s} \frac{B'_{is} B'_{js}}{p+r_s}. \tag{3.12}$$

‡ We follow here the procedure developed in the author's paper 'Theory of stress-strain relations in anisotropic viscoelasticity and relaxation phenomena', *J. appl. Phys.* **25**, 1385–91 (1954).

This expression is immediately reducible to a constant term, a term proportional to p and a sum of partial fractions $C_s/(p+r_s)$.

It is of interest, however, to derive a slightly different expression for Z_{ij} that corresponds to a spring-dashpot analogue model as introduced by the author in the general context of irreversible thermodynamics.‡ This particular form of Z_{ij} is

$$Z_{ij} = \sum^{s} \frac{p}{p+r_s} D_{ij}^{(s)} + D_{ij} + D'_{ij} p. \tag{3.13}$$

This result is obtained by substituting into expression (3.12) the values (3.9) for B'_{is} and B'_{js} and putting

$$\begin{aligned} D_{ij} &= a_{ij} - \sum^{s} \frac{a'_{is} a'_{js}}{r_s}, & D'_{ij} &= b_{ij} - \sum^{s} b'_{is} b'_{js}, \\ D_{ij}^{(s)} &= \psi_i^{(s)} \psi_j^{(s)}, & \psi_i^{(s)} &= \frac{a'_{is}}{r_s^{\frac{1}{2}}} - b'_{is} r_s^{\frac{1}{2}}. \end{aligned} \tag{3.14}$$

An important remark at this point concerns the possibility that some values of r_s may be zero. However, this does not cause the values of D_{ij} and $\psi_i^{(s)}$ to become infinite, because whenever r_s is zero the corresponding values of a'_{is} must also vanish. In order to show this we consider the value (3.4) of V. If $r_s = 0$ with $a'_{is} \neq 0$, we can make the term $a'_{is} q_i \xi_s$ negative and arbitrarily large. Hence V could become negative, which is contrary to a basic property of the thermal potential.

Note that there may be multiple characteristic roots of the sub-system. Hence we may proceed as in the similar case for the value (2.10) of $C_{ij}^{(s)}$ by introducing the coefficient

$$D_{ij}^{(s)} = \sum^{s'} \psi_i^{(s+s')} \psi_j^{(s+s')}, \tag{3.15}$$

where $\sum^{s'}$ is extended to the multiplicity corresponding to a particular multiple root r_s.

As in the case of expression (2.11), we conclude that the quadratic form $\sum^{ij} D_{ij}^{(s)} z_i z_j$ is non-negative. That $\sum^{ij} D_{ij} z_i z_j$ and $\sum^{ij} D'_{ij} z_i z_j$ are also non-negative quadratic forms can be seen as follows. Equations (2.1) may be written

$$\sum^{\nu} (a_{\mu\nu} + p b_{\mu\nu}) q_\nu = Q_\mu, \tag{3.15 a}$$

where μ and ν vary from 1 to n. We multiply these equations by q_μ and derive

$$\sum^{\mu\nu} a_{\mu\nu} q_\mu q_\nu + p \sum^{\mu\nu} b_{\mu\nu} q_\mu q_\nu = \sum^{\mu} Q_\mu q_\mu. \tag{3.15 b}$$

Since $Q_{k+1} = Q_{k+2} = \dots = Q_n = 0$, we may write

$$\sum^{\mu} Q_\mu q_\mu = \sum^{i} Q_i q_i = \sum^{ij} Z_{ij} q_i q_j. \tag{3.15 c}$$

‡ See reference on p. 47.

For $p = 0$ we obtain the following result:

$$\sum^{\mu\nu} a_{\mu\nu} q_\mu q_\nu = \sum^{ij} D_{ij} q_i q_j; \tag{3.15d}$$

and for $p = \infty$,

$$\sum^{\mu\nu} b_{\mu\nu} q_\mu q_\nu = \sum^{ij} D'_{ij} q_i q_j. \tag{3.15e}$$

The left sides of equations (3.15d) and (3.15e) are non-negative since they correspond to the thermal potential and the dissipation function. Hence the quadratic forms on the right side are also non-negative.

4. FOURIER AND LAPLACE TRANSFORMS

The transient response of a thermal system to arbitrary time-dependent forces $Q_i(t)$ applied at the instant $t = 0$ may be derived from the admittance matrix of elements $A_{ij}(p)$. There are three closely related methods by which this can be done, using either Fourier transforms, Laplace transforms, or operational rules. Consider the *Fourier transform*,

$$G_j(i\omega) = \int_{-\infty}^{+\infty} e^{-i\omega t} Q_j(t)\, dt. \tag{4.1}$$

If we assume $Q_j(t) = 0$ for $t < 0$, it may be written

$$G_j(i\omega) = \int_0^{\infty} e^{-i\omega t} Q_j(t)\, dt. \tag{4.2}$$

The inverse transform is

$$Q_j(t) = \frac{1}{2\pi} \int_{-\infty}^{+\infty} e^{i\omega t} G_j(i\omega)\, d\omega. \tag{4.3}$$

With the variable

$$p = i\omega, \tag{4.4}$$

it may be written

$$Q_j(t) = \frac{1}{2\pi i} \int_{-i\infty}^{+i\infty} e^{pt} G_j(p)\, dp. \tag{4.5}$$

This expression represents the force $Q_j(t)$ as a superposition of harmonic components, each proportional to the factor $\exp(i\omega t)$. A particular solution of the differential equations

$$\sum^{k} a_{jk} q_k + \sum^{k} b_{jk} \dot{q}_k = Q_j(t),$$

which govern the thermal system, is therefore obtained by introducing the admittance $A_{kj}(p)$ discussed in Section 2. This solution is

$$q_k(t) = \frac{1}{2\pi i} \int_{-i\infty}^{+i\infty} e^{pt} A_{kj}(p) G_j(p)\, dp. \tag{4.6}$$

To simplify the writing, we have assumed here that all other forces except $Q_j(t)$ are zero. The general case is, of course, obtained by superposition of all such solutions.

We must still insure that the solution (4.6) satisfies the initial conditions. We shall assume that initially the system is quiescent. This is expressed by the condition

$$q_k(t) = 0 \quad \text{for } t < 0. \tag{4.7}$$

In general, it is easy to verify this condition by deforming the path of integration in the integral (4.6) in such a way that its value vanishes for $t < 0$. That this is possible in most problems is due to the fact that $A_{kj}(p)$ and $G_j(p)$, considered as analytic functions of p, have no singularities in the half-plane corresponding to a positive real part of p. That this is the case for $A_{kj}(p)$ can be seen from the partial fraction expansion (2.11). Whether the property is valid for $G_j(p)$ must be verified by evaluating the integral (4.2). However, for most functions encountered in physical problems it is found to be valid.

This is best illustrated by a simple example. Consider a force $Q_j(t)$ which is a Heaviside function

$$Q_j(t) = 1(t). \tag{4.7 a}$$

This is also called a unit step function and is defined as follows:

$$\begin{aligned} 1(t) &= 0 \quad \text{for } t < 0, \\ 1(t) &= 1 \quad \text{for } t > 0. \end{aligned} \tag{4.7 b}$$

This Heaviside function may be represented by a Fourier transform as

$$1(t) = \frac{1}{2\pi i} \int_{c-i\infty}^{c+i\infty} \frac{e^{pt}}{p}\, dp. \tag{4.7 c}$$

It is easily verified by a standard procedure that this integral represents the discontinuous function $1(t)$. The value of c is arbitrarily small but positive. Hence the line of integration is arbitrarily near the imaginary axis and is in the half-plane where p has a positive real part. If the admittance element $A_{kj}(p)$ relates the response q_k and the force Q_j, the response $q_k = \alpha_k(t)$ to the unit step (4.7 c) is

$$\alpha_k(t) = \frac{1}{2\pi i} \int_{c-i\infty}^{c+i\infty} e^{pt} \frac{A_{kj}(p)}{p}\, dp. \tag{4.7 d}$$

Since $A_{kj}(p)$ has no singularities in the half-plane where the real part of p is positive, it is verified, as in the case of equation (4.7 c), that $\alpha_k(t) = 0$ for $t < 0$. Hence the initial condition that the system is quiescent is satisfied.

Note that once $\alpha_k(t)$ has been determined the response $q_k(t)$ to an arbitrary force $Q_j(t)$ is obtained by superposition as a *Duhamel integral*:

$$q_k(t) = \int_0^t \alpha_k(t-t')\, dQ_j(t'). \tag{4.7 e}$$

This expression must be interpreted as a Stieltjes integration.

We may write equation (4.2) in the form

$$G_j(p) = \int_0^\infty e^{-pt}Q_j(t)\,dt, \tag{4.8}$$

where p is now real and positive. The function $G_j(p)$ is then the *Laplace transform* of $Q_j(t)$. When the Laplace transform $G_j(p)$ is given, relation (4.8) is an integral equation for $Q_j(t)$. Under very broad conditions its solution is unique. Relation (4.8) is also known as *Carson's integral equation*.

A fundamental theorem for Laplace transforms concerns their product. Consider the Laplace transforms $\Lambda_1(p)$, $\Lambda_2(p)$ of two functions $f_1(t)$ and $f_2(t)$ defined for $t > 0$. Hence

$$\Lambda_1(p) = \int_0^\infty e^{-pt}f_1(t)\,dt, \qquad \Lambda_2(p) = \int_0^\infty e^{-pt}f_2(t)\,dt. \tag{4.9}$$

The theorem states that the product of these Laplace transforms is

$$\Lambda_1(p)\Lambda_2(p) = \int_0^\infty e^{-pt}f_3(t)\,dt, \tag{4.10}$$

where
$$f_3(t) = \int_0^t f_1(t-t')f_2(t')\,dt' = \int_0^t f_2(t-t')f_1(t')\,dt'. \tag{4.11}$$

Therefore the Laplace transform of the *convolution* f_3 of two functions f_1 and f_2 is obtained by forming the product of their Laplace transforms. This property provides the basis of an operational algebra which may be translated into operational rules.‡

5. OPERATIONAL RULES

Instead of writing explicitly the integral transforms, it is possible to manipulate directly the differential and integral operators as algebraic quantities.

For example, the differential equations are written in symbolic form by using the symbol

$$p = \frac{d}{dt} \tag{5.1}$$

‡ For a modern treatment of operational calculus see, for example, Balth. van der Pol and H. Bremmer, *Operational calculus based on the two-sided Laplace transform*. Cambridge University Press (1950).

as a differential operator. The equations

$$\sum^{k} (a_{jk}+pb_{jk})q_k = Q_j \tag{5.2}$$

may be interpreted as differential equations or as relations between Laplace transforms. In other words, the variables q_k and Q_j may be interpreted in two ways. They may represent functions of time, in which case p is said to *operate* on q_k as a time differential. They may also represent their Laplace transforms.

The solution of equations (5.2) as given by relations (2.2) is

$$q_k = \sum^{j} A_{kj}(p)Q_j. \tag{5.3}$$

Again this may be interpreted as a relation between Laplace transforms. However, if q_k and Q_j are functions of time, then $A_{kj}(p)$ may be considered as an operator acting upon the functions Q_j. Some basic operational rules are as follows:

$$\begin{aligned} pf(t) &= \frac{d}{dt}f(t), \\ \frac{1}{p}f(t) &= \int_0^t f(t')\,dt', \\ \frac{1}{p+a}f(t) &= e^{-at}\int_0^t e^{at'}f(t')\,dt'. \end{aligned} \tag{5.4}$$

In order to avoid ambiguities we must assume that the functions $f(t)$ start with the value $f(0) = 0$. However, it may jump to a discontinuous value at the instant $t > 0$ as close as desired to the origin. Application of operational rules to such cases, by considering discontinuities to be defined as a limiting case of continuous functions, is straightforward with the use of generalized functions. These operational rules may be derived from the integral equation (4.8).‡ The third operational rule (5.4) is seen to correspond to a particular solution of the differential equation

$$(p+a)z = f(t) \tag{5.5}$$

with initial condition $z(0) = 0$. Let us apply the operational rules to the case where $f(t)$ is equal to the Heaviside function

$$f(t) = 1(t). \tag{5.6}$$

‡ See, for example, Th. von Kármán and M. A. Biot, *Mathematical methods in engineering*. McGraw-Hill, New York (1940).

Expressions (5.4) become

$$
\begin{aligned}
p1(t) &= \delta(t) \quad \text{(Dirac function)},\\
\frac{1}{p}\,1(t) &= t, \qquad\qquad (5.7)\\
\frac{1}{p+a}\,1(t) &= \frac{1}{a}(1-e^{-at}).
\end{aligned}
$$

Note that by the use of generalized functions the operators p and $1/p$ become commutative. These operational rules may also be applied to the admittance operator $A_{kj}(p)$. Suppose only the force $Q_j(t)$ is applied as a known function of time. Equation (5.3) reduces to

$$q_k = A_{kj}(p)Q_j(t), \tag{5.8}$$

where $A_{kj}(p)$ is now operating on the function $Q_j(t)$. In order to interpret this operator we go back to its expansion (2.11) in partial fractions. This expansion is

$$A_{kj}(p) = \sum^{s} \frac{C_{kj}^{(s)}}{\lambda_s+p}. \tag{5.9}$$

Hence, applying the third operational rule (5.4), equation (5.8) is written as

$$q_k = \sum^{s} C_{kj}^{(s)}\, e^{-\lambda_s t} \int_0^t e^{\lambda_s t'} Q_j(t')\, dt'. \tag{5.10}$$

We note that the terms for which $\lambda_s = 0$ are represented by the operator $1/p$, which is simply an integration

$$\frac{1}{p}\,Q_j(t) = \int_0^t Q_j(t')\, dt'. \tag{5.11}$$

As another example, consider the impedance operator. Expanded in partial fractions, its value, as given by equation (3.13), is

$$Z_{kj}(p) = \sum^{s} \frac{p}{p+r_s}\, D_{kj}^{(s)} + D_{kj} + D'_{kj}\, p. \tag{5.12}$$

A generalized coordinate $q_j(t)$, which is a given function of time, generates a thermal force

$$Q_k = Z_{kj}(p)q_j(t). \tag{5.13}$$

The operational interpretation of this expression yields

$$Q_k = \sum^{s} D_{kj}^{(s)}\, e^{-r_s t} \int_0^t e^{r_s t'} \frac{dq_j(t')}{dt'}\, dt' + D_{kj}\, q_j(t) + D'_{kj} \frac{d}{dt} q_j(t). \tag{5.14}$$

6. OPERATOR–VARIATIONAL PRINCIPLE

Let us go back to equations (2.2) and (2.4) of Chapter 1 and replace the time differential by the operator p. They become

$$c\theta = -\operatorname{div}\mathbf{H}, \qquad \operatorname{grad}\theta+\frac{p}{k}\mathbf{H} = 0. \tag{6.1}$$

In these equations we may interpret the variables θ and $\mathbf{H}$ as representing their Laplace transforms. In this case p becomes an algebraic quantity. We may then solve equations (6.1) using again Laplace transforms for the space boundary conditions.

We may obviously write the variational principle (2.9) of Chapter 1 with Laplace transforms. Hence

$$\delta V+p\iiint\limits_{\tau}\frac{1}{k}\mathbf{H}.\delta\mathbf{H}\,d\tau = -\iint\limits_{A}\theta\mathbf{n}.\delta\mathbf{H}\,dA. \tag{6.2}$$

As before, $$V = \tfrac{1}{2}\iiint\limits_{\tau}c\theta^2\,d\tau, \qquad \delta V = \iiint\limits_{\tau}c\theta\,\delta\theta\,d\tau. \tag{6.3}$$

By introducing the quadratic form

$$\mathscr{D} = \frac{1}{2}\iiint\limits_{\tau}\frac{1}{k}\mathbf{H}^2\,d\tau \tag{6.4}$$

and putting $$\delta Q = -\iint\limits_{A}\theta\mathbf{n}.\delta\mathbf{H}\,dA \tag{6.5}$$

the variational principle (6.2) becomes

$$\delta V+p\,\delta\mathscr{D} = \delta Q. \tag{6.6}$$

This is a particular case of the operator-variational principle developed by the author in the more general context of linear thermodynamics.‡ The operational results are easily extended to the case of anisotropic thermal conductivity.

The meaning of the operator-variational principle (6.6) is very broad. It provides a compact formulation of several different types of manipulations. In fact it may be interpreted in three different ways.

1. Operational interpretation

In this case we use generalized coordinates related linearly to the field

‡ 'Thermoelasticity and irreversible thermodynamics', *J. appl. Phys.* **27**, 240–53 (1956), and 'Variational and Lagrangian methods in viscoelasticity', *Deformation and flow of solids* (IUTAM Colloquium, Madrid, 1955), pp. 251–63. Springer, Berlin (1956).

H. As in equation (3.1) of Chapter 2, we write

$$\mathbf{H} = \sum^{i} \mathbf{H}^{(i)}(x, y, z) q_i. \tag{6.7}$$

The variational principle (6.6) leads to the equation

$$\frac{\partial}{\partial q_i}(V + p\mathscr{D}) = Q_i, \tag{6.8}$$

where

$$V = \tfrac{1}{2} \sum^{ij} a_{ij} q_i q_j,$$

$$\mathscr{D} = \tfrac{1}{2} \sum^{ij} b_{ij} q_i q_j, \tag{6.9}$$

$$Q_i = -\iint_A \theta \mathbf{n} . \mathbf{H}^{(i)}\, dA.$$

In explicit form, equations (6.8) are

$$\sum^{j} (a_{ij} + p b_{ij}) q_j = Q_i. \tag{6.10}$$

With $p = d/dt$, these are the differential equations of the thermal system. What has been done here is to manipulate p as an algebraic symbol in the variational principle and then replace p by d/dt in the final linear equations.

2. Algebraic interpretation

As already pointed out in connection with the integral equation (4.8), the Laplace transform as a numerical function of real positive values of p uniquely determines the corresponding function of time. In the Laplace transform domain, we may solve the problem for every positive real value of p by using the variational principle (6.6). Since p is now a real positive quantity, the solution may be obtained in numerical form, or may be part numerical and part algebraic. Space boundary conditions must, of course, be satisfied with δQ a given function of p. The approximate variational solution thus determined as a function of p, is then expressed as a function of time by transform inversion. In practice, this may conveniently be accomplished by using simple analytical approximations for the functions of p.

3. Convolution interpretation

It is well known that the product of two Laplace transforms represents a convolution of the corresponding functions as indicated by equation

(4.10). Hence, in the variational principle (6.2), we may put

$$\theta\,\delta\theta = \int_0^t \theta(t-t')\,\delta\theta(t')\,dt',$$

$$\mathbf{H}p\,\delta\mathbf{H} = \int_0^t \mathbf{H}(t-t')\,\delta\dot{\mathbf{H}}(t')\,dt', \tag{6.11}$$

$$\theta\,\delta\mathbf{H} = \int_0^t \theta(t-t')\,\delta\mathbf{H}(t')\,dt'.$$

The quantities θ and $\mathbf{H}$ are now functions of both the time and the space coordinates and are related by the constraint $c\theta = -\operatorname{div}\mathbf{H}$. Substitution of expressions (6.11) into equation (6.2) yields a variational principle for the unknown field $\mathbf{H}$ as a function of both time and space. Applications of the convolution interpretation of operator-variational principles have been discussed by Schapery‡ and Gurtin.§

7. INTERCONNECTION PRINCIPLE

We shall consider a thermal system to be divided into a certain number of domains connected by common boundaries. Each domain constitutes a sub-system that may be analysed separately. In particular, we may determine the thermal impedance of each sub-system. We will show that in this case it is possible to derive the operational or integro-differential equations for the total system by a variational principle. The principle is entirely analogous to the principle of virtual work in mechanics.‖

Let the various domains be designated by a number s (Fig. 3.1). The thermal impedance of the sub-system corresponding to this domain is designated by $Z_{ij}^{(s)}$. As before, when considering the impedance we must define the associated external coordinates. In the present case they are defined in the following way.

Consider the boundary AB common to the domains s and $s+1$ (Fig. 3.1). We choose a distribution of heat displacement whose normal component at the interface AB is expressed as $H_p\,q_i$, where H_p is a function only of the location on the interface. This defines a generalized

‡ R. A. Schapery, 'On the time dependence of viscoelastic variational solutions', *Q. appl. Math.* **22**, 207–15 (1964).

§ M. E. Gurtin, 'Variational principles for linear initial value problems', ibid. 252–6 (1964).

‖ The existence of such a principle in the broader context of irreversible thermodynamics was outlined in the author's paper 'Variational principles in irreversible thermodynamics with application to viscoelasticity', *Phys. Rev.* **97**, 1463–9 (1955).

coordinate q_i for the interface AB. It represents a common external coordinate for the two adjacent systems s and $s+1$. In general, we may define a certain number of such external coordinates for any boundary

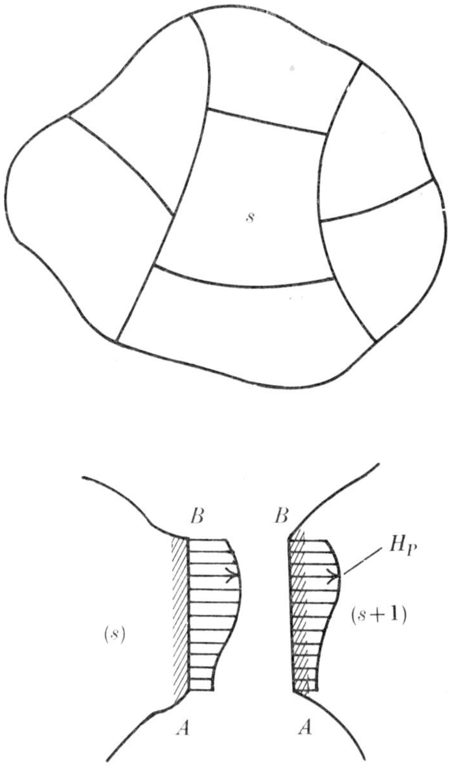

FIG. 3.1. Division into sub-system domains and definition of generalized coordinates at an interface AB.

between two adjacent domains. For each domain s we may write the equations

$$Q_i = \sum^j Z_{ij}^{(s)} q_j, \tag{7.1}$$

where q_j are the generalized coordinates for the boundary of this domain and Q_i the associated thermal forces at this boundary. Because $Z_{ij}^{(s)} = Z_{ji}^{(s)}$, equation (7.1) leads to the variational relation

$$\sum^i Q_i \delta q_i = \delta Z^{(s)}, \tag{7.2}$$

where

$$Z^{(s)} = \tfrac{1}{2} \sum^{ij} Z_{ij}^{(s)} q_i q_j. \tag{7.3}$$

The summations here are extended to the variables at the boundary of the domain s.

We now take the sum of equations (7.2) for all domains and write

$$\sum^{i} Q_i \,\delta q_i = \delta Z, \tag{7.4}$$

where

$$Z = \sum^{s} Z^{(s)}. \tag{7.5}$$

The significance of this result lies in the interpretation of $Q_i\,\delta q_i$. The summation $\sum^{i}$ is extended to the variables at the boundary of the total system and at the interfaces. However, at the interfaces the terms may be grouped in pairs that cancel out. Consider, for example, the coordinate q_i at the interface AB. For the domain s the corresponding thermal force is $Q_i^{(s)}$ and for the domain $s+1$ it is $Q_i^{(s+1)}$. The outward normal vectors of these two domains are in opposite directions, while the applied temperatures are the same. Hence

$$Q_i^{(s)} = -Q_i^{(s+1)}, \tag{7.6}$$

and we derive

$$(Q_i^{(s)}+Q_i^{(s+1)})\,\delta q_i = 0. \tag{7.7}$$

Therefore all forces at the interfaces disappear from the summation in the variational principle (7.4). As can be seen, this variational principle leads to the equations

$$\frac{\partial Z}{\partial q_i} = Q_i, \qquad \frac{\partial Z}{\partial q_j} = 0. \tag{7.8}$$

In the first group of equations the thermal forces Q_i are those applied to the external boundaries of the total system, and q_i are the generalized coordinates at these boundaries. In the second group the coordinates q_j are those at the interfaces of the sub-systems.

Because of the operational nature of Z equations (7.8) constitute an integro-differential system for the unknowns q_i and q_j.

Complementary form of the interconnection principle

We may choose the heat displacements at common boundaries to be of opposite signs. The generalized forces Q_i are then equal at these boundaries. Again we may write equations such as (7.1) for each domain s. We then solve these equations for q_i:

$$q_i = \sum^{j} A_{ij}^{(s)}\, Q_j, \tag{7.9}$$

with an operational matrix $[A_{ij}^{(s)}]$ which is the inverse of the matrix $[Z_{ij}^{(s)}]$. Complementary quadratic forms may be defined as

$$A^{(s)} = \tfrac{1}{2} \sum^{ij} A_{ij}^{(s)}\, Q_i\, Q_j \tag{7.10}$$

and

$$A = \sum^{s} A^{(s)}. \tag{7.11}$$

With these definitions the following complementary variational principle is verified with arbitrary variation:

$$\sum^{i} q_i \delta Q_i = \delta A. \tag{7.12}$$

It is a consequence of equations (7.9). The summation $\sum^{i} q_i \delta Q_i$ contains only the variables q_i at the boundary of the total system. The variables q_i at the common boundaries of the sub-systems drop out because they appear in pairs of opposite sign. From the variational principle (7.12) we derive the operational equations

$$\frac{\partial A}{\partial Q_j} = q_j, \qquad \frac{\partial A}{\partial Q_i} = 0. \tag{7.13}$$

In these relations, in contrast to equations (7.8), the temperatures at interconnecting boundaries are treated as unknowns while q_j are the heat displacement variables at the boundary of the total system. In general, they will correspond to integro-differential equations.

Application to finite element methods

The numerical analysis of a transient field may be carried out by dividing the domain into finite elements. For example, in a two-dimensional field the elements may constitute a triangular network. The size of the elements may vary, being small in areas of steep gradients near boundaries, and larger in areas of smoother behaviour. The choice of a network will be dictated by criteria of accuracy and computing economy.

We then evaluate the $Z_{ij}^{(s)}$ operators for a typical cell. A system of equations for the heat displacements at interconnecting boundaries is obtained by applying the interconnection principle (7.4) or by the equivalent procedure of equating the thermal forces at these boundaries with suitable change in sign.

In the complementary formulation we evaluate the operators of equations (7.9). This leads to equations (7.13) for the unknown thermal forces. These unknowns may be defined in terms of temperatures at the vertices of a network.

8. CONTINUOUS SPECTRUM

We shall apply the operational method to the problem already considered in Section 7 of Chapter 2. In addition to providing an illustrative example, the treatment leads to the concept of continuous relaxation spectrum.

In this problem a slab of thickness l is heated suddenly to a constant temperature θ_0 at the boundary $x = 0$ at the time $t = 0$, while the boundary $x = l$ is maintained at zero temperature. The heat displacement is represented as

$$H = q_0 + \sum^n q_n \cos\frac{n\pi x}{l}; \tag{8.1}$$

the corresponding temperature is

$$\theta = \frac{\pi}{cl}\sum^n nq_n \sin\frac{n\pi x}{l}. \tag{8.2}$$

Since the problem is one-dimensional we evaluate V and $\mathscr{D}$ by using as volume of integration a cylinder of unit cross-section, parallel to x. The values (6.3) and (6.4) for V and D become in the present case

$$V = \frac{1}{4}\frac{\pi^2}{cl}\sum^n n^2 q_n^2, \qquad \mathscr{D} = \frac{l}{2k}q_0^2 + \frac{1}{4}\frac{l}{k}\sum^n q_n^2. \tag{8.3}$$

If θ_0 is the temperature at $x = 0$, the values of Q_0 and Q_n are

$$Q_0 = Q_n = \theta_0. \tag{8.4}$$

With these values the operational equations (6.8) become

$$(\lambda_n + p)q_n = \frac{2k}{l}\theta_0, \qquad pq_0 = \frac{k}{l}\theta_0, \tag{8.5}$$

with

$$\lambda_n = \frac{k\pi^2 n^2}{cl^2}. \tag{8.6}$$

The operational solution of equations (8.5) is

$$q_n = \frac{1}{p+\lambda_n}\frac{2k}{l}\theta_0, \qquad q_0 = \frac{k}{pl}\theta_0. \tag{8.7}$$

The right side of these equations represents an operation on the function θ_0. In this case θ_0 is the constant temperature suddenly applied at $t = 0$. Hence we must replace θ_0 by $\theta_0 1(t)$. According to the operational rules (5.7), equations (8.7) are interpreted as

$$q_n = \frac{2k\theta_0}{l\lambda_n}(1 - e^{-\lambda_n t}), \qquad q_0 = \frac{k}{l}\theta_0 t. \tag{8.8}$$

This result coincides with equations (7.14) of Chapter 2.

We may also express the thermal admittance at $x = 0$. With this value of x the heat displacement (8.1) becomes

$$H = q_0 + \sum^n q_n. \tag{8.9}$$

We introduce the values (8.7) for q_0 and q_n and derive

$$H = A(p)\theta_0, \tag{8.10}$$

where

$$A(p) = \frac{k}{pl} + \sum^{n} \frac{2k}{l(p+\lambda_n)} \tag{8.11}$$

is the thermal admittance at the boundary $x = 0$. It is a particular case of the general expression (2.11) for the thermal admittance. Note that the value (2.11) may include terms of the form $1/p$ corresponding to zero values of λ_s.

By considering the limiting case of a slab of infinite thickness l we are led to the concept of a continuous relaxation spectrum. In the limit the summation in equation (8.11) is replaced by an integral. In order to show this we evaluate the increment

$$\Delta\surd\lambda = (\lambda_{n+1})^{\frac{1}{2}} - (\lambda_n)^{\frac{1}{2}} = \frac{\pi}{l}\left(\frac{k}{c}\right)^{\frac{1}{2}}. \tag{8.12}$$

Hence

$$\frac{1}{l} = \frac{1}{\pi}\left(\frac{c}{k}\right)^{\frac{1}{2}} \Delta\surd\lambda. \tag{8.13}$$

For small values of $1/l$ the increment $\Delta\surd\lambda$ may be replaced by a differential

$$\Delta\surd\lambda = \frac{d\lambda}{2\surd\lambda}. \tag{8.14}$$

In the limit, the summation of equation (8.11) may be replaced by the integral

$$A(p) = \frac{(kc)^{\frac{1}{2}}}{\pi} \int_0^\infty \frac{1}{p+\lambda} \frac{d\lambda}{\surd\lambda}. \tag{8.15}$$

The admittance is thus represented by means of a continuous spectrum of relaxation constants λ with a density distribution $1/\surd\lambda$.

By performing the integration in equation (8.15) we derive

$$A(p) = \left(\frac{kc}{p}\right)^{\frac{1}{2}}. \tag{8.16}$$

This is the thermal admittance of the half-space.

Note that this result could have been derived directly by writing the equation of heat diffusion in operational form

$$k\frac{d^2\theta}{dx^2} = pc\theta. \tag{8.16 a}$$

The operational solution of this equation (with boundary conditions $\theta = \theta_0$ at $x = 0$ and $\theta = 0$ at $x = \infty$) is

$$\theta = \theta_0 \exp\left\{-x\left(\frac{cp}{k}\right)^{\frac{1}{2}}\right\}. \tag{8.16 b}$$

The heat displacement at $x = 0$ is

$$H = -\frac{k}{p}\left(\frac{\partial\theta}{\partial x}\right)_{x=0} = \left(\frac{kc}{p}\right)^{\frac{1}{2}}\theta_0. \tag{8.16 c}$$

Hence H/θ_0 coincides with the value (8.16) of $A(p)$.

It should be noted that the summation signs in expressions (2.11) and (3.13) for the thermal admittance and impedance may also be replaced, either in part or completely, by integrations with spectral density distributions, if some boundaries of the thermal system are at infinity. We may also substitute integrals for the summations as an approximate representation.

CHAPTER FOUR

ASSOCIATED FIELDS

1. INTRODUCTION

In the foregoing analysis the thermal field is described by means of heat displacement vectors. This leads to a formulation that has the advantage of satisfying rigorously conservation of energy. At the same time, use of the heat displacement vectors makes it possible to introduce the powerful method of virtual work, thus providing a unified treatment analogous to the methods of classical mechanics. However, the use of a vector field representation generally introduces more unknowns than in the scalar field representation by means of temperature. This is particularly true in two- and three-dimensional problems.

In order to avoid this difficulty, a method was devised that combines the advantages of using a scalar field while retaining energy conservation and the applicability of the virtual work principle. We have called it the method of *associated fields*.‡ It is derived from the existence of *ignorable coordinates* in the same sense as in classical mechanics. What is accomplished by the use of associated fields may be stated as follows.

(1) It decouples the ignorable coordinates and the temperature field, thereby reducing considerably the number of unknowns in two- and three-dimensional problems while respecting energy conservation.

(2) It introduces in the coordinates themselves features that contain basic properties of the physical system such as reciprocity. The coordinates therefore represent already a partial solution of the problem. This further reduces the number of coordinates required.

(3) Problems are solved by a two-step process, one being the determination of the associated field from a principle of minimum dissipation followed by the integration of the differential equations for the unknown coordinates representing the temperature field.

(4) In steady-state problems, temperatures in one part of the system are determined independently from those in other regions. They are

‡ This method and its various corollaries as presented in this chapter was developed in detail in the author's paper, 'Further development of new methods in heat flow analysis', *J. Aerospace Sci.* **26**, 367–81 (1959).

obtained immediately for any distribution of externally applied temperatures without having to repeat the calculations for each case.

The existence of ignorable coordinates that may be decoupled from the temperature field is derived in Section 2. This leads to the definition of the associated field as a vector field defined by a given temperature field. As shown in Section 3, the associated field may be obtained from the temperature field by applying a principle of minimum dissipation. An alternative derivation of the associated field in Section 4 leads to an analogue model and it is shown that this model also obeys the principle of minimum dissipation. The relation of associated fields to Green's function is discussed in Section 5.

From the standpoint of normal coordinates, the existence of decoupled ignorable coordinates is a consequence of orthogonality properties. This viewpoint is developed in Section 6. It leads to the concept of *ignorable subspace* as the infinite set of normal coordinates with vanishing characteristic value. It is a case of infinite degeneracy. As indicated, the evaluation of associated fields for normal coordinates is immediate. An application to quasi-steady solutions is discussed which provides a very simple method of separating the temperature field into a sequence of steady state solutions and a superposition of relaxation modes.

A specific example of the use of associated fields is treated in Section 7. A structure composed of a web and flanges is analysed by introducing the concept of 'effective flange width'.

Attention should be called to the method of elimination of ignorable coordinates in mechanics which is based on matrix algebra and orthogonality properties. For example, in the vibration analysis of a free body there are six degenerate modes of zero frequency corresponding to rigid rotations and translations. They may be eliminated by using orthogonality properties. In practice such methods are applicable if the number of ignorable coordinates is not large. In the case of infinite degeneracy as described in Section 6, it is preferable to accomplish the elimination of the ignorable coordinates by applying the principle of minimum dissipation.

2. IGNORABLE COORDINATES AND ASSOCIATED FIELDS

We may write the heat displacement field in the form

$$\mathbf{H} = \sum^{i} \mathbf{\Theta}_i q_i + \sum^{l} \mathbf{F}_l f_l, \tag{2.1}$$

where $$\boldsymbol{\Theta}_i = \boldsymbol{\Theta}_i(x, y, z), \qquad \mathbf{F}_l = \mathbf{F}_l(x, y, z) \tag{2.2}$$

are fixed vector fields. The latter field is assumed to be divergence-free, hence

$$\operatorname{div} \mathbf{F}_l = 0. \tag{2.3}$$

The generalized coordinates have thus been separated into two groups. The system is defined by ν coordinates q_i and k coordinates f_l. Together they constitute the $n = \nu + k$ coordinates, which adequately describe the behaviour of the thermal system. The temperature field θ is obtained from $\mathbf{H}$ by the relation

$$\theta = -\frac{1}{c} \operatorname{div} \mathbf{H}. \tag{2.4}$$

Substituting the value (2.1) for $\mathbf{H}$ and taking into account condition (2.3), we obtain

$$\theta = \sum^{i} \theta_i q_i, \tag{2.5}$$

where $$\theta_i = -\frac{1}{c} \operatorname{div} \boldsymbol{\Theta}_i. \tag{2.6}$$

In these expressions the heat capacity $c(x, y, z)$ may be a function of the coordinates.

Under these conditions the thermal potential depends only on the ν coordinates q_i. It is expressed by the quadratic form

$$V = \tfrac{1}{2} \sum^{ij} a_{ij} q_i q_j. \tag{2.7}$$

In the summation, i and j assume all values from 1 to ν.

On the other hand, the dissipation function will contain all of the $\nu + k$ coordinates q_i and f_l. It is written

$$D = \tfrac{1}{2} \sum^{ij} b_{ij} \dot{q}_i \dot{q}_j + \sum^{il} b'_{il} \dot{q}_i \dot{f}_l + \tfrac{1}{2} \sum^{lm} b''_{lm} \dot{f}_l \dot{f}_m. \tag{2.8}$$

In the summation, i and j assume all values from 1 to ν, while l and m assume all values from 1 to k. The quadratic form D is positive-definite. It is obtained by applying expressions (5.18) of Chapter 1, or (2.12) of Chapter 2, for either isotropic or anisotropic thermal conductivity. We assume here that the thermal conductivity, $k(x, y, z)$ or $k_{ij}(x, y, z)$, may depend on the coordinates and is independent of the time.

From V and D we derive the differential equations for the time history of the system. They are

$$\frac{\partial V}{\partial q_i} + \frac{\partial D}{\partial \dot{q}_i} = Q_i, \qquad \frac{\partial D}{\partial \dot{f}_l} = Q_l. \tag{2.9}$$

In these equations Q_i and Q_l are the thermal forces corresponding respectively to the coordinates q_i and f_l. Substitution of expressions (2.7) and (2.8) into equations (2.9) yields the explicit form

$$\begin{aligned} \sum^{j} a_{ij} q_j + \sum^{j} b_{ij} \dot{q}_j + \sum^{l} b'_{il} \dot{f}_l &= Q_i, \\ \sum^{i} b'_{il} \dot{q}_i + \sum^{m} b''_{lm} \dot{f}_m &= Q_l. \end{aligned} \tag{2.10}$$

In this system the k coordinates f_l do not appear in the value (2.7) of the thermal potential. Such a system will be said to possess k ignorable coordinates.

This terminology introduced by the author in the broader context of dissipative phenomena‡ is the same as in classical dynamics where it refers to the case of a system with $\nu+k$ degrees of freedom where the potential energy contains only ν coordinates. The dynamical equations may then be reduced to ν equations with ν unknowns and the k remaining coordinates are designated as ignorable. The same procedure is applicable for the thermal system governed by equations (2.10). We will show that the equations may be reduced to a system where the variables q_i and new variables f'_l are uncoupled.

Let the system be defined by the same ν coordinates q_i and in addition by a new set of k coordinates f'_l. In terms of these new variables the old coordinates are written as

$$f_l = f'_l + \sum^{j} \alpha_{lj} q_j, \tag{2.11}$$

where α_{lj} are $k \times \nu$ coefficients to be determined. By substituting the values (2.11) of f_l into the dissipation function (2.8), we obtain

$$D = \tfrac{1}{2} \sum^{ij} \mathscr{B}_{ij} \dot{q}_i \dot{q}_j + \sum^{il} \mathscr{B}'_{il} \dot{q}_i \dot{f}'_l + \tfrac{1}{2} \sum^{lm} b''_{lm} \dot{f}'_l \dot{f}'_m. \tag{2.12}$$

The new coefficients are

$$\begin{aligned} \mathscr{B}_{ij} &= b_{ij} + \sum^{l} (b'_{il} \alpha_{lj} + b'_{jl} \alpha_{li}) + \sum^{lm} b''_{lm} \alpha_{li} \alpha_{mj}, \\ \mathscr{B}'_{il} &= b'_{ij} + \sum^{m} b''_{lm} \alpha_{mi}. \end{aligned} \tag{2.13}$$

The variables, q_i and f'_l, are uncoupled if

$$\mathscr{B}'_{il} = b'_{il} + \sum^{m} b''_{lm} \alpha_{mi} = 0. \tag{2.14}$$

For every index i these relations constitute a set of k equations for the k unknowns α_{mi}. Since D is positive-definite the partial quadratic form $\sum^{lm} b''_{lm} \dot{f}_l \dot{f}_m$ in expression (2.8) is also positive-definite. Hence the determinant of the coefficients b''_{lm} is different from zero. In that case

‡ M. A. Biot, 'Thermoelasticity and irreversible thermodynamics', *J. appl. Phys.* **27**, 240–53 (1956).

equations (2.14) have a unique solution for α_{mi}. The uncoupled equations are written as

$$\sum^{j} a_{ij} q_j + \sum^{j} \mathscr{B}_{ij} \dot{q}_j = Q'_i, \qquad \sum^{m} b''_{lm} \dot{f}'_m = Q'_l. \tag{2.15}$$

Hence we have derived a system of two independent sets of differential equations, one for the group of ν variables q_i, the other for the group of k variables f'_l. The variables f'_l are the *ignorable coordinates.*

The new forces, Q'_i, and Q'_l are determined by the principle of virtual work (see p. 9):

$$\sum^{i} Q'_i \delta q_i + \sum^{l} Q'_l \delta f'_l = \sum^{i} Q_i \delta q_i + \sum^{l} Q_l \delta f_l. \tag{2.16}$$

By substituting for δf_l the values derived from equations (2.11) and identifying the coefficients of δq_i and $\delta f'_l$ on both sides of equation (2.16), we obtain

$$Q'_i = Q_i + \sum^{l} Q_l \alpha_{li}, \qquad Q'_l = Q_l. \tag{2.17}$$

The significance of these results is further clarified as follows.

Suppose we have evaluated the coefficients α_{li} by solving equations (2.14). Let us substitute the values (2.11) of f_l into expression (2.1) for the heat displacement field. We obtain

$$\mathbf{H} = \sum^{i} \mathbf{\Theta}'_i q_i + \sum^{l} \mathbf{F}_l f'_l \tag{2.18}$$

with

$$\mathbf{\Theta}'_i = \mathbf{\Theta}_i + \sum^{l} \mathbf{F}_l \alpha_{li}. \tag{2.19}$$

If we are interested only in the temperature we may drop the terms containing the ignorable coordinates f'_l. In this case we start with the temperature

$$\theta = \sum^{i} \theta_i q_i \tag{2.20}$$

and an *associated* heat displacement,

$$\mathbf{H}' = \sum^{i} \mathbf{\Theta}'_i q_i. \tag{2.21}$$

The component vector fields $\mathbf{\Theta}'_i$ may also be called associated with the corresponding scalar fields θ_i. Note that the following relation is satisfied:

$$\theta_i = -\frac{1}{c} \operatorname{div} \mathbf{\Theta}'_i. \tag{2.22}$$

If we drop the terms containing the ignorable coordinates the dissipation function (2.12) is reduced to

$$D' = \tfrac{1}{2} \sum^{ij} \mathscr{B}_{ij} \dot{q}_i \dot{q}_j. \tag{2.23}$$

This leads for the coordinates q_i to the Lagrangian equations

$$\frac{\partial V}{\partial q_i}+\frac{\partial D'}{\partial \dot{q}_i} = Q'_i, \tag{2.24}$$

which contain only the ν unknowns q_i.

As can be seen, the temperatures may be obtained independently from the heat flow by starting with a superposition of fixed temperature configuration θ_i, as given by expression (2.20), and evaluating for each field θ_i an *associated vector field* $\mathbf{\Theta}'_i$.

3. MINIMUM DISSIPATION PRINCIPLE FOR ASSOCIATED FIELDS

We will show that for a given temperature field θ_i the associated field $\mathbf{\Theta}'_i$ may be evaluated by applying a principle of minimum dissipation.

Consider that all the non-ignorable coordinates are zero except q_i. The heat displacement (2.1) is reduced to

$$\mathbf{H} = \mathbf{\Theta}_i q_i + \sum^{l} \mathbf{F}_l f_l. \tag{3.1}$$

Let us choose the k coordinates f_l to be proportional to q_i and write

$$f_l = \beta_{li} q_i. \tag{3.2}$$

The k coefficients β_{li} are yet undetermined. The heat displacement (3.1) becomes

$$\mathbf{H} = \Big(\mathbf{\Theta}_i + \sum^{l} \mathbf{F}_l \beta_{li}\Big) q_i. \tag{3.3}$$

The time derivative of $\mathbf{H}$ is

$$\dot{\mathbf{H}} = \Big(\mathbf{\Theta}_i + \sum^{l} \mathbf{F}_l \beta_{li}\Big) \dot{q}_i. \tag{3.4}$$

Hence the dissipation function due to the variation of q_i alone is

$$D = \Big(\tfrac{1}{2} b_{ii} + \sum^{l} b'_{il} \beta_{li} + \tfrac{1}{2} \sum^{lm} b''_{lm} \beta_{li} \beta_{mi}\Big) \dot{q}_i^2. \tag{3.5}$$

It is obtained by substituting $\dot{f}_l = \beta_{li} \dot{q}_i$ into expression (2.8).

Let us now choose the coefficients β_{li} in such a way that the dissipation D be a minimum. This requires the following equations to be satisfied:

$$\frac{\partial D}{\partial \beta_{li}} = 0. \tag{3.6}$$

Hence

$$b'_{il} + \sum^{m} b''_{lm} \beta_{mi} = 0. \tag{3.7}$$

But these relations are precisely the same as equations (2.14), which determine the coefficients α_{mi}. Hence

$$\beta_{li} = \alpha_{li}. \tag{3.8}$$

Therefore, by choosing the coefficients β_{li} in such a way that they minimize the dissipation function, the bracketed expression in equation (3.4) becomes the associated field (2.19),

$$\mathbf{\Theta}'_i = \mathbf{\Theta}_i + \sum^{l} \mathbf{F}_l \alpha_{li}. \tag{3.9}$$

Analogue model for the associated field

These results may be expressed in a slightly different form. We start with a given temperature field

$$\theta = \theta_i q_i \tag{3.10}$$

and a heat displacement field

$$\mathbf{H} = \mathbf{H}_i q_i. \tag{3.11}$$

The field $\mathbf{H}_i$ is not determined by θ_i. However, it is constrained by the condition

$$c\theta_i = -\operatorname{div} \mathbf{H}_i. \tag{3.12}$$

This constraint may also be written

$$w = \operatorname{div} \dot{\mathbf{H}} \tag{3.13}$$

with

$$w = -c\dot{\theta} = -c\theta_i \dot{q}_i. \tag{3.14}$$

Equation (3.13) provides an analogue model such that the rate of heat flow $\dot{\mathbf{H}}$ is the same as that due to a distributed heat source with a rate of heat generation equal to w per unit volume. We may also say that the heat flow is due to *heat sinks* of magnitude $c\dot{\theta} = -w$. We put $\dot{q}_i = 1$ and write

$$w = -c\theta_i = \operatorname{div} \mathbf{H}_i, \qquad \dot{\mathbf{H}} = \mathbf{H}_i. \tag{3.15}$$

Under these conditions $\mathbf{H}_i$ represents the rate of heat flow due to the heat sinks $c\theta_i = -w$. If we impose the additional condition that the dissipation function due to $\dot{\mathbf{H}} = \mathbf{H}_i$ is a minimum, then $\mathbf{H}_i$ becomes the associated field

$$\mathbf{H}_i = \mathbf{\Theta}'_i. \tag{3.16}$$

In the next section we shall derive the same result by a different procedure and we will show that this minimum condition is equivalent to the equations for steady state heat flow due to the sinks $c\theta_i$ combined with the boundary condition that the outside adiabatic temperature θ_a is zero.

4. ALTERNATIVE FORMULATION FOR ASSOCIATED FIELDS

The nature and properties of the associated field are further clarified by using a somewhat different approach.

For simplicity, consider first the case of a system with isotropic thermal conductivity $k(x,y,z)$ function of the coordinates. The heat capacity $c(x,y,z)$ is also a function of the coordinates.

The boundary A of the volume τ of the medium involves linear heat transfer properties characterized by a coefficient $K(x,y,z)$, which may depend on the location.

Consider a scalar field

$$\psi = \psi(q_1, q_2,..., q_\nu, x, y, z), \tag{4.1}$$

which is a function of ν generalized coordinates q_ν. In addition, consider a vector field

$$F = \mathbf{F}(f_1, f_2,..., f_k, x, y, z), \tag{4.2}$$

which depends on k generalized coordinates f_l. We impose the condition that this field be divergence-free:

$$\operatorname{div} \mathbf{F} = 0. \tag{4.3}$$

We write the heat displacement field $\mathbf{H}$ in the form

$$\mathbf{H} = -k \operatorname{grad} \psi + \mathbf{F}. \tag{4.4}$$

This expression represents the unknown field $\mathbf{H}$ as a function of the $\nu+k$ coordinates q_i and f_l. The temperature θ is related to the vector field by the relation

$$c\theta = -\operatorname{div} \mathbf{H}. \tag{4.5}$$

By substituting the value (4.4) we obtain

$$c\theta = \operatorname{div}(k \operatorname{grad} \psi). \tag{4.6}$$

Hence the temperature field depends only on q_i. We will show that it is possible to choose ψ in such a way that in the differential equations for the generalized coordinates the fields ψ and $\mathbf{F}$ become uncoupled.

The differential equations for q_i and f_l are written as

$$\frac{\partial V}{\partial q_i} + \frac{\partial D}{\partial \dot{q}_i} = Q_i, \qquad \frac{\partial D}{\partial \dot{f}_l} = Q_l. \tag{4.7}$$

Since θ depends only on q_i, the coordinates f_l do not appear in the thermal potential V. In addition, the generalized thermal forces Q_i involve only q_i. Hence, in equations (4.7) coupling is due to the dissipation function D. It is evaluated by applying expression (2.13) of Chapter 2, which includes the boundary dissipation. It may be written as the sum of three terms:

$$D = D_q + D_{qf} + D_f. \tag{4.8}$$

The term D_q depends on q_i and $\dot{q}_i$, while D_f depends on f_l and $\dot{f}_l$. Hence coupling arises only through the term

$$D_{qf} = -\iiint_{\tau} \dot{\mathbf{F}} \operatorname{grad} \dot{\psi}\, d\tau - \iint_{A} \frac{k}{K} \dot{F}_n \operatorname{grad}_n \dot{\psi}\, dA, \tag{4.9}$$

where $\dot{F}_n$ and $\operatorname{grad}_n \dot{\psi}$ denote the normal components of the vectors at the boundary. They are projected on the outward normal direction.

Integration by parts of the volume integral in expression (4.9), taking into account the property $\operatorname{div} \dot{\mathbf{F}} = 0$, yields the surface integral

$$D_{qf} = -\iint_{A} \dot{F}_n \left(\dot{\psi} + \frac{k}{K} \operatorname{grad}_n \dot{\psi}\right) dA.$$

This expression vanishes if at the boundary the following condition is satisfied

$$K\psi + k \operatorname{grad}_n \psi = 0. \tag{4.10}$$

Hence the fields ψ and $\mathbf{F}$ are uncoupled if ψ verifies the boundary condition (4.10).

This result may be interpreted as follows. The temperature field

$$\theta = \theta(q_1, q_2, \ldots, q_\nu, x, y, z) \tag{4.11}$$

may be chosen as the unknown. For any given value of θ we may integrate the partial differential equation (4.6) with the boundary condition (4.10). For every field θ this determines uniquely a function ψ and a vector field

$$\mathbf{\Theta}' = -k \operatorname{grad} \psi. \tag{4.12}$$

This is the heat displacement *associated with the temperature field* (4.11).

Analogue model for associated fields

These results lead to a more precise formulation of the analogue model derived in the previous section. Consider a temperature field $\theta(q_1, q_2, \ldots, q_\nu, x, y, z)$ for given values of q_i; it is a function only of x, y, z. We distribute heat sinks throughout the volume with a rate of heat disappearance per unit volume equal to

$$-w = c\theta. \tag{4.13}$$

We also impose the condition that the heat flows from an outside domain of zero adiabatic temperature ($\theta_a = 0$). We denote by ψ the steady-state temperature distribution due to these sinks. It is obtained by solving the equation

$$-w = \operatorname{div}(k \operatorname{grad} \psi), \tag{4.14}$$

with the boundary condition

$$K\psi + k \operatorname{grad}_n \psi = 0. \tag{4.15}$$

The value of the temperature thus obtained is the same as the value of ψ derived from equations (4.6) and (4.10). Hence the field associated with θ is

$$\mathbf{\Theta}' = -k \operatorname{grad} \psi. \tag{4.16}$$

This vector represents the rate of heat flow in the analogue heat-sink model.

If the temperature is expressed as a superposition of fields

$$\theta = \sum^{i} \theta_i q_i, \tag{4.17}$$

where $\theta_i(x, y, z)$ are given scalar configurations, the corresponding heat displacement is

$$\mathbf{H} = \sum^{i} \mathbf{\Theta}'_i q_i. \tag{4.18}$$

Each associated field $\mathbf{\Theta}'_i$ is derived from θ_i by an analogue model. It is given by the rate of heat flow due to steady sinks $-w_i = c\theta_i$ and zero adiabatic temperature at the boundary. Note that if the boundary does not include a heat transfer layer ($K = \infty$) the adiabatic temperature is equal to the temperature of the solid itself at the surface. In this case a zero temperature is maintained at the surface of the solid in the analogue model.

Associated field derived by minimizing the dissipation

It is not necessary to evaluate the temperature ψ of the analogue model. The rate of flow $\mathbf{\Theta}'$ due to the sinks may be obtained directly by applying the same minimum principle already formulated in the previous section. The alternate derivation of this principle, which follows, demonstrates at the same time the identity of the two analogue models described here and in the previous section. Consider the dissipation due to the rate of heat flow $\mathbf{\Theta}'$. It is written

$$D = \frac{1}{2} \iiint_{\tau} \frac{1}{k} \mathbf{\Theta}'^2 \, d\tau + \frac{1}{2} \iint_{A} \frac{1}{K} \Theta_n'^2 \, dA. \tag{4.19}$$

Since it is a steady flow generated by heat sinks $-w$ the vector field $\mathbf{\Theta}'$ must satisfy the condition

$$w = \operatorname{div} \mathbf{\Theta}'. \tag{4.20}$$

The minimum of D when varying $\boldsymbol{\Theta}$ under the constraint (4.20) is obtained by minimizing

$$D' = \iiint \left[\frac{1}{2k}\boldsymbol{\Theta}'^2 - \Lambda(\operatorname{div}\boldsymbol{\Theta}' - w)\right] d\tau + \frac{1}{2}\iint_A \frac{1}{K}\Theta_n'^2\, dA, \tag{4.21}$$

where Λ is a Lagrangian multiplier. This yields Euler's equation,

$$\frac{1}{k}\boldsymbol{\Theta}' + \operatorname{grad}\Lambda = 0 \tag{4.22}$$

and the boundary condition

$$\frac{1}{K}\Theta_n' - \Lambda = 0. \tag{4.23}$$

By taking into account relation (4.20), we derive

$$-w = \operatorname{div}(k \operatorname{grad}\Lambda) \tag{4.24}$$

with the boundary condition

$$K\Lambda + k \operatorname{grad}_n \Lambda = 0. \tag{4.25}$$

Comparing these results with equations (4.14) and (4.15), we see that

$$\Lambda = \psi. \tag{4.26}$$

Equation (4.22) then yields

$$\boldsymbol{\Theta}' = -k \operatorname{grad}\psi. \tag{4.27}$$

Hence the field $\boldsymbol{\Theta}'$ derived by minimizing the dissipation (4.19) is the same as the associated field given by equation (4.12).

Further generalization

In the foregoing derivation we have assumed that expressions (4.1) and (4.11) for θ and ψ do not contain the time explicitly. However, such an assumption is not required. We may start with a temperature field,

$$\theta = \theta(q_1, q_2,..., q_\nu, x, y, z, t), \tag{4.27 a}$$

which contains the time explicitly. The associated field may then be derived as follows. At a given instant t and for given values of q_i we distribute sinks $-w = c\theta$ in the volume, assuming zero adiabatic temperature at the boundary. The steady state temperature ψ under these conditions yields the associated field,

$$\boldsymbol{\Theta}'(q_1, q_2,..., q_\nu, x, y, z, t) = -k \operatorname{grad}\psi. \tag{4.27 b}$$

The results obtained in this section have been derived in the context of isotropic thermal conductivity. They are readily extended to the anisotropic case. The method of determination of the associated field by considering an analogue model with distributed heat sinks along with the principle of minimum dissipation, remains valid for anisotropic conductivity.

5. RELATION TO GREEN'S FUNCTION

We consider again a temperature field which we write as

$$\theta(x) = \theta(q_1, q_2, \ldots, q_\nu, x, y, z).$$

For given values of q_i this is a known function of x, y, z. In the notation $\theta(x)$, the variable x represents in abbreviated form the three coordinates x, y, z. The vector field associated with $\theta(x)$ may be expressed in terms of Green's function. In order to simplify the writing, and without loss of generality, we shall consider a system with isotropic conductivity. Let us write Dirac's function in abbreviated form,

$$\delta(x, x') = \delta(x-x', y-y', z-z'). \tag{5.1}$$

The notation x and x' stand respectively for the coordinates x, y, z and x', y', z'. We define Green's function $g(x, x')$ by requiring that it satisfies the equation

$$\operatorname{div}(k \operatorname{grad} g) = c(x')\, \delta(x, x') \tag{5.2}$$

with the boundary condition

$$Kg + k \operatorname{grad}_n g = 0. \tag{5.3}$$

In these expressions $c(x') = c(x', y', z')$ denotes the heat capacity and the operations div and grad are performed on the variables x, y, z.

A known property of Dirac's function is expressed by the relation

$$c(x)\theta(x) = \iiint_\tau c(x')\theta(x')\, \delta(x, x')\, dx', \tag{5.4}$$

where dx' stands for dx', dy', dz'. Hence equation (4.6) for ψ may be written as

$$\operatorname{div}(k \operatorname{grad} \psi) = \iiint_\tau c(x')\theta(x')\, \delta(x, x')\, dx'. \tag{5.5}$$

Because g satisfies relation (5.2), a solution ψ of equation (5.5) is obtained by superposition as

$$\psi = \iiint_\tau g(x, x')\theta(x')\, dx'. \tag{5.6}$$

Condition (5.3) for g implies that this value of ψ also satisfies the boundary condition (4.10). With this value of ψ expression (4.12) for the field associated with $\theta(x)$ becomes

$$\boldsymbol{\Theta}'(x) = -k \iiint_\tau (\operatorname{grad} g)\theta(x')\, dx'. \tag{5.7}$$

If the temperature field is expressed in the form

$$\theta(x) = \sum^{i} \theta_i(x) q_i, \tag{5.8}$$

the corresponding heat displacement is

$$\mathbf{H}(x) = \sum^{i} \mathbf{\Theta}'_i(x) q_i. \tag{5.9}$$

To each temperature configuration $\theta_i(x)$ is associated a vector field $\mathbf{\Theta}'_i(x)$. Each associated field is obtained from $\theta_i(x)$ in terms of Green's function by applying equation (5.7). We write

$$\mathbf{\Theta}'_i(x) = -k \iiint_{\tau} (\operatorname{grad} g) \theta_i(x')\, dx'. \tag{5.10}$$

Green's function is also applicable when $\theta(x)$ is chosen to be an explicit function of the time t. In this case θ is expressed as

$$\theta(x) = \theta(q_1, q_2, \ldots, q_\nu, x, y, z, t). \tag{5.11}$$

The associated field is obtained by substituting this value of $\theta(x)$ into equation (5.7).

Application to steady-state solutions

The formulation of associated fields by means of Green's function provides a connection between the method developed here and certain classical results in the theory of partial differential equations.

Consider the steady state generated by time-independent adiabatic temperatures θ_a applied at the boundary. The steady state is represented by the temperature field, which is reached asymptotically after sufficient time under the constant thermal forces. This steady-state solution is obtained readily by using associated fields.

To show this, the temperature field is written as a linear superposition of fields expressed by equation (5.8) with the corresponding associated field (5.9) Because we are using associated fields the steady state requires that q_i be constant. Hence $\dot{q}_i = 0$ and the dissipation function vanishes from the formulation. By putting $D = 0$ the Lagrangian equations for the steady state are reduced to

$$\frac{\partial V}{\partial q_i} = Q_i. \tag{5.12}$$

With the value (5.8) for θ, the thermal potential is

$$V = \tfrac{1}{2} \iiint_{\tau} c\theta^2\, dx = \tfrac{1}{2} \sum^{ij} a_{ij} q_i q_j, \tag{5.13}$$

where
$$a_{ij} = \iiint_{\tau} c(x) \theta_i(x) \theta_j(x)\, dx. \tag{5.14}$$

According to equation (3.12) of Chapter 2, the thermal forces are

$$Q_i = -\iint_A \theta_a \mathbf{\Theta}'_i . \mathbf{n}\, dA. \tag{5.15}$$

The surface integral is extended to the boundary A of the volume τ.

We shall now introduce a very particular choice of generalized coordinates by dividing the volume τ into small cells of volume τ_i. Each cell is located at the point of coordinates x_i. Consider the cell τ_i. We define a function of x such that

$$\begin{aligned} \theta_i(x) &= 1 \quad \text{inside the cell,} \\ \theta_i(x) &= 0 \quad \text{outside the cell.} \end{aligned} \tag{5.16}$$

We may then write the temperature field in the form

$$\theta(x) = \sum^{i} \theta_i(x)\theta(x_i), \tag{5.17}$$

where $\theta(x_i)$ is the local temperature at the point x_i. Comparing with expression (5.8) we see that the local temperature plays the role of the generalized coordinates

$$\theta(x_i) = q_i. \tag{5.18}$$

By introducing the values (5.16) into expression (5.10), we obtain the following associated fields:

$$\mathbf{\Theta}'_i(x) = -k\{\operatorname{grad} g(x, x_i)\}\tau_i. \tag{5.19}$$

Evaluation of a_{ij} from equations (5.14) yields the thermal potential,

$$V = \tfrac{1}{2}\sum^{i} c(x_i)\tau_i q_i^2 = \tfrac{1}{2}\sum^{i} c(x_i)\theta^2(x_i)\tau_i. \tag{5.20}$$

The thermal force (5.15) is

$$Q_i = \tau_i \iint_A k\theta_a\, \mathbf{n}.\{\operatorname{grad} g(x, x_i)\}\, dA. \tag{5.21}$$

With these values of V and Q_i the Lagrangian equations (5.12) become

$$c(x_i)\theta(x_i) = \iint_A k\theta_a\, \mathbf{n}.\{\operatorname{grad} g(x, x_i)\}\, dA. \tag{5.22}$$

Hence the local temperature $\theta(x_i)$ is determined directly by means of the adiabatic boundary temperature θ_a in terms of Green's function $g(x, x_i)$. Equation (5.22) coincides with a classical result in the theory of Green's function.

It should be pointed out that a direct evaluation of the local steady-state temperature does not require that Green's function be known. This can be seen by writing equations (5.12) with the values (5.15) and

(5.20) for Q_i and V respectively. It becomes

$$c(x_i)\theta(x_i) = -\iint_A \theta_a \mathbf{\Theta}'_i . \mathbf{n}\, dA. \tag{5.23}$$

In this expression the local temperature $\theta(x_i)$ of the cell τ_i is obtained by a simple quadrature once we know the associated field $\mathbf{\Theta}'_i$. This field may be evaluated by minimizing the dissipation of the flow due to heat sinks of magnitude $c(x)\theta_i(x)$. According to the definition (5.16) of $\theta_i(x)$, this amounts to distributing sinks of magnitude $c(x_i)$ uniformly in the cell τ_i.

6. ASSOCIATED FIELDS AND NORMAL COORDINATES

In Chapter 2 we discussed the formulation of thermal flow by means of normal coordinates. According to equations (5.17) of that chapter, the thermal potential and the dissipation function in terms of generalized coordinates ξ_s are expressed as follows:

$$V = \tfrac{1}{2}\sum^{s} \lambda_s \xi_s^2, \qquad D = \tfrac{1}{2}\sum^{s} \dot{\xi}_s^2. \tag{6.1}$$

A complete representation of the thermal system includes heat displacement fields of vanishing divergence. These fields do not affect the temperature, and the corresponding coordinates do not appear in the thermal potential. In the normal coordinate representation these fields correspond to zero values of the characteristic values λ_s.

Let us assume that the ν values $\lambda_1, \lambda_2,..., \lambda_\nu$ are different from zero, while the k remaining values vanish. We write

$$\lambda_s = 0, \tag{6.2}$$

where $s = \nu+1, \nu+2,..., \nu+k$. Expressions (6.1) become

$$V = \tfrac{1}{2}\sum_{s=1}^{\nu} \lambda_s \xi_s^2, \qquad D = \tfrac{1}{2}\sum_{s=1}^{\nu} \dot{\xi}_s^2 + \tfrac{1}{2}\sum_{s=\nu+1}^{\nu+k} \dot{\xi}_s^2. \tag{6.3}$$

The ignorable coordinates in this case are $\xi_{\nu+1}, \xi_{\nu+2},..., \xi_{\nu+k}$. In expressions (6.3) they are already uncoupled from the non-ignorable coordinates $\xi_1, \xi_2,..., \xi_\nu$. In addition, they are uncoupled from each other.

The existence of ignorable coordinates as discussed in Section 2 may be considered as a direct consequence of expressions (6.3). This can be verified as follows. Let us introduce the linear transformation

$$\xi_s = \sum_{i=1}^{\nu} \beta_{si}\, q_i, \tag{6.4}$$

where $s = 1, 2,..., \nu$. Similarly, we introduce the transformation

$$\xi_s = \sum_{l=1}^{k} \gamma_{sl} f'_l, \tag{6.5}$$

where $s = \nu+1, \nu+2,..., \nu+k$. By substituting the values (6.4) and (6.5) into expressions (6.3) we derive V and D in the form

$$V = \tfrac{1}{2} \sum^{ij} a_{ij} q_i q_j, \qquad D = \tfrac{1}{2} \sum^{ij} \mathscr{B}_{ij} \dot{q}_i \dot{q}_j + \tfrac{1}{2} \sum^{lm} b''_{lm} \dot{f}'_l \dot{f}'_m. \tag{6.6}$$

In the summations, i and j vary from 1 to ν, while l and m vary from 1 to k. The variables f'_l do not appear in the value of V. They represent the ignorable coordinates. Moreover, all coupling terms between q_i and f'_l are zero. These results coincide with those of Section 2, where the existence of uncoupled coordinates was already derived.

Infinite degeneracy—ignorable subspace

As can be seen from the foregoing discussion, the possibility of decoupling the ignorable and non-ignorable coordinates may be considered as a consequence of the properties of normal coordinates. We may look upon the ignorable coordinates as a linear combination of degenerate relaxation modes with a vanishing characteristic root λ_s. Because of orthogonality properties, this linear combination is uncoupled to the relaxation modes corresponding to the non-ignorable coordinates and for which λ_s is different from zero.

The uncoupled equations for the ignorable coordinates are derived from expression (6.6) for D. They are

$$\sum^{m} b''_{lm} \dot{f}'_m = F_l, \tag{6.7}$$

where F_l is the thermal force conjugate to the coordinate f'_l. From the mathematical viewpoint, a complete description of the field corresponding to the ignorable coordinates requires an infinite number of such coordinates. Actually, we are dealing here with a functional space of all fields $\mathbf{H}$ having the property $\operatorname{div}\mathbf{H} = 0$. It may be called an *ignorable subspace*. We may also represent this subspace by an infinite number of normal coordinates ξ_s with characteristic roots all equal to zero. Hence there are an infinite number of multiple zero roots. The ignorable subspace may be said to have *infinite degeneracy*. The rate of heat flow $\dot{\mathbf{H}}$ in the ignorable subspace is given by the values of $\dot{f}'_m$. Equations (6.7) show that they are determined uniquely by the thermal forces F_l. Note that the values F_l may be time-dependent since they are determined by the adiabatic temperature θ_a applied at the boundary.

However, in the ignorable subspace, the flow rate $\dot{\mathbf{H}}$ at any particular instant is the same as in the steady state obtained by applying a constant adiabatic boundary temperature equal to θ_a at that particular instant. In other words, the solution in the ignorable subspace is represented by a continuous sequence of steady-state fields.

Instead of a representation by generalized coordinates we may use a continuum representation. Consider the case of isotropic thermal conductivity. The flow rate in the ignorable subspace is expressed as

$$\dot{\mathbf{H}} = -k\,\mathrm{grad}\,\theta, \tag{6.8}$$

where θ satisfies the partial differential equation

$$\mathrm{div}(k\,\mathrm{grad}\,\theta) = 0 \tag{6.9}$$

and the boundary condition

$$K(\theta-\theta_a)+k\,\mathrm{grad}_n\,\theta = 0 \tag{6.10}$$

(K = boundary heat transfer coefficient). As can be seen, these equations determine the temperature θ under steady-state conditions with a given adiabatic temperature θ_a at the boundary. Similar equations corresponding to an instantaneous steady state are obtained for the case of anisotropic thermal conductivity.‡

Associated fields for normal coordinates

We now consider the case where the temperature field is represented by means of normal coordinates. We write

$$\theta = \sum^{s} \theta_s \xi_s, \tag{6.11}$$

where
$$\theta_s = \theta_s(x, y, z) \tag{6.12}$$

are normalized temperature fields defining characteristic solutions as in equation (5.11) of Chapter 2. The heat displacement is written

$$\mathbf{H} = \sum^{s} \mathbf{\Theta}'_s \xi_s, \tag{6.13}$$

where $\mathbf{\Theta}'_s(x, y, z)$ is the vector field associated with θ_s. In the case of normal coordinates these associated fields are easily derived as follows.

Let us assume isotropic conductivity. Since θ_s represents a characteristic solution, the relaxation mode

$$\theta = \theta_s e^{-\lambda_s t} \tag{6.14}$$

is a solution of the heat conduction equation

$$\mathrm{div}(k\,\mathrm{grad}\,\theta) = c\frac{\partial\theta}{\partial t}. \tag{6.15}$$

Hence
$$\mathrm{div}(k\,\mathrm{grad}\,\theta_s) = -c\lambda_s \theta_s. \tag{6.16}$$

‡ See also Section 4 in the Appendix.

The boundary condition satisfied by the relaxation mode is

$$K\theta_s + k\,\mathrm{grad}_n\,\theta_s = 0. \tag{6.17}$$

In Section 4 it has been shown that the associated field is given by the expression

$$\Theta'_s = -k\,\mathrm{grad}\,\psi, \tag{6.18}$$

where ψ satisfies the equation

$$\mathrm{div}(k\,\mathrm{grad}\,\psi) = c\theta_s \tag{6.19}$$

and the boundary condition

$$K\psi + k\,\mathrm{grad}_n\,\psi = 0. \tag{6.20}$$

The solution ψ of these equations is readily obtained by taking into account the identities (6.16) and (6.17). We derive

$$\psi = -\frac{1}{\lambda_s}\theta_s. \tag{6.21}$$

Hence the associated field is

$$\boldsymbol{\Theta}'_s = \frac{k}{\lambda_s}\,\mathrm{grad}\,\theta_s. \tag{6.22}$$

For anisotropic conductivity a similar derivation yields the following expression for the associated field:

$$\Theta'_{is} = \frac{1}{\lambda_s}\sum^{j} k_{ij}\frac{\partial\theta_s}{\partial x_j}, \tag{6.23}$$

where Θ'_{is} are the components of $\boldsymbol{\Theta}'_s$ and x_j designates the coordinates x, y, z.

Application to quasi-steady flow

In Chapter 2, Section 6, we have formulated a quasi-steady solution by considering the continuous sequence of instantaneous steady states to which is added a correction expressed by generalized coordinates. A similar quasi-steady solution may be obtained by using normal coordinates.‡ We represent the temperature field by means of normal coordinates

$$\theta = \sum^{s} \theta_s q_s, \tag{6.24}$$

with the associated field

$$\mathbf{H} = \sum^{s} \boldsymbol{\Theta}'_s q_s, \tag{6.25}$$

where $\boldsymbol{\Theta}'_s$ is given by expression (6.22). The thermal potential and dissipation function are

$$V = \tfrac{1}{2}\sum^{s} \lambda_s q_s^2, \qquad D = \tfrac{1}{2}\sum^{s} \dot{q}_s^2. \tag{6.26}$$

The Lagrangian equations are

$$\lambda_s q_s + \dot{q}_s = Q_s, \tag{6.27}$$

‡ For further details see M. A. Biot, 'Further developments of new methods in heat flow analysis' *J. Aerospace Sci.* **26**, 367–81 (1959).

where
$$Q_s = -\iint_A \theta_a \mathbf{\Theta}'_s . \mathbf{n}\, dA. \tag{6.28}$$

This expression involves the boundary surface integral of the adiabatic temperature θ_a. We separate the solution into an instantaneous steady state q_s^* and a correction q_s^+. Hence
$$q_s = q_s^* + q_s^+. \tag{6.29}$$
The instantaneous steady state q_s^* is the equilibrium state that would be obtained for adiabatic temperatures θ_a if they were time-independent. It is obtained by putting $\dot{q}_s = 0$ in equations (6.27). We derive
$$q_s^* = \frac{Q_s}{\lambda_s}. \tag{6.30}$$
Substitution of expression (6.29) into equations (6.27) yields
$$\lambda_s q_s^+ + \dot{q}_s^+ = -\frac{\dot{Q}_s}{\lambda_s}, \tag{6.31}$$
where
$$\dot{Q}_s = -\iint_A \dot{\theta}_a \mathbf{\Theta}'_s . \mathbf{n}\, dA. \tag{6.32}$$
The temperature field is now
$$\theta = \theta^* + \theta^+, \tag{6.33}$$
where θ^* is given by the instantaneous steady state and the correction θ^+ is expressed as
$$\theta^+ = \sum^s \theta_s q_s^+. \tag{6.34}$$
The generalized coordinates q_s^+ are obtained by solving equations (6.31). Note that the quantities on the right side of equations (6.31) play the role of fictitious thermal forces proportional to the time derivative $\dot{\theta}_a$ of the adiabatic temperature. The steady-state field θ^* may be determined directly without using generalized coordinates by solving the partial differential equation for steady-state temperature with given boundary value for θ_a. The example treated in Section 7 of Chapter 2 provides an illustration of this separation into a steady-state solution and a correction expressed by normal coordinates.

7. EXAMPLE OF ASSOCIATED FIELDS

In order to illustrate the use of associated fields in problems of thermal conduction, we shall consider a simple structure‡ whose cross-section is

‡ This example is taken from the author's paper, 'Thermodynamics and heat flow analysis by Lagrangian methods', *Proceedings of the Seventh Anglo-American Aeronautical Conference*, pp. 418–31. Institute of the Aeronautical Sciences, New York (1959).

shown in Fig. 4.1. It is composed of two flanges of thickness a connected by a web of thickness $2a_1$. The material is assumed homogeneous with constant values of the thermal conductivity k and the heat capacity c. The problem is treated as a two-dimensional temperature distribution in the plane of the figure. The structure is heated by the application of adiabatic temperatures on the outside surface of the flanges and the

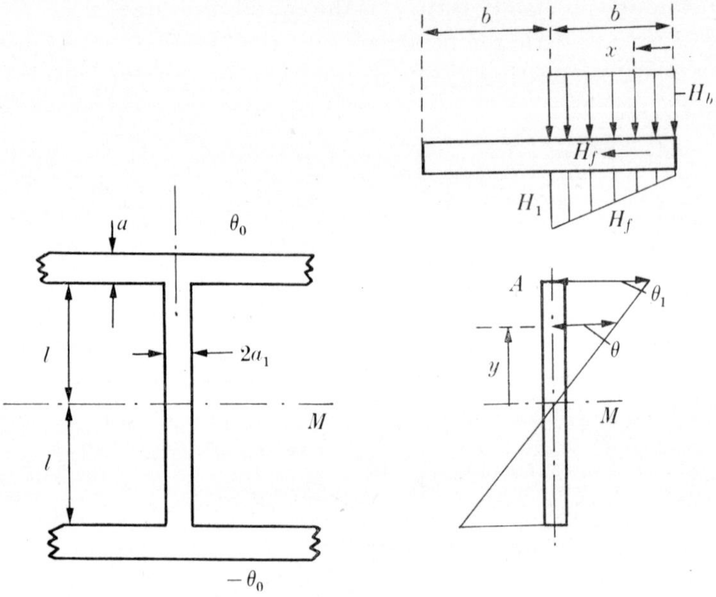

FIG. 4.1. Associated field for heat flow in a structure composed of a web and two flanges.

heating occurs by means of a surface heat-transfer coefficient K. It is assumed here that the use of a surface heat-transfer coefficient is justified. If the surface heating is achieved through contact with a moving fluid the assumption is probably justified if the fluid motion is normal to the figure.

However, if the fluid motion is in the plane of the figure, strong reservations must be made regarding the use of a surface heat-transfer coefficient, and other methods must be used. They will be discussed in detail in Chapters 6 and 7.

We shall evaluate the steady state temperature in the web due to constant adiabatic temperatures along the outer surfaces of the flanges. A uniform temperature is applied at the top flange and a different one at the bottom flange. Because of the symmetry of the structure, the temperature field may be separated into a symmetric and an antisymmetric distribution about the mid-point of the web. We assume that the adiabatic temperatures on the top and bottom flanges are θ_0 and $-\theta_0$. Then only the antisymmetric web temperature will be produced. We shall solve the antisymmetric steady-state case for a constant value of θ_0 by applying the method of associated fields.

The temperature along the web has a linear distribution

$$\theta = \frac{y}{l}\theta_1. \tag{7.1}$$

The unknown temperature θ_1 at the joint A will be considered as the generalized coordinate. The field associated with the temperature distribution (7.1) in the web may be obtained by applying the analogue model described in Section 4. In this model we distribute thermal sinks along the web with an intensity $c\theta$ per unit volume. This will produce a flow H_f through the flange and a corresponding flow H_b normal to the flange. We shall assume that H_b is distributed uniformly over a width $2b$ of the flange. The flow H_f through the flange is then linearly distributed. The length b plays the role of an 'effective width', and is determined by a principle of minimum dissipation. We evaluate the dissipation in the half-width b due to the surface heat transfer and the flow H_f in the flange itself. According to expression (4.19) this dissipation may be written

$$D_1 = \frac{a}{2k}\int\limits_0^b H_f^2\,dx + \frac{b}{2K}H_b^2, \tag{7.2}$$

where
$$H_f = H_1\frac{x}{b}, \qquad H_b = a\frac{H_1}{b}. \tag{7.3}$$

The value of H_f at the joint A is represented by H_1. For a given value of H_1 expression (7.2) is a function of b. The value of b that minimizes D_1 is

$$b = \sqrt{\left(\frac{3ak}{K}\right)}. \tag{7.4}$$

With this value of b we write

$$D_1 = \frac{a^2H_1^2}{Kb}. \tag{7.5}$$

The flow H_w in the web is determined by the flow H_1 and the distributed sinks. We find

$$H_w = \frac{aH_1}{a_1} - c\int\limits_y^l \theta\,dy, \tag{7.6}$$

where $2l$ is the distance between flanges.

The dissipation in the half-thickness a_1 of the web between points A and M is

$$D_2 = \frac{a_1}{2k}\int\limits_0^l H_w\,dy. \tag{7.7}$$

The total dissipation is $$D = D_1+D_2, \tag{7.8}$$

where D_1 has the value (7.5). The value of H_1 is obtained by minimizing D as a function of H_1. This determines completely all components of the associated field in terms of the coordinate θ_1. In particular, we find

$$H_b = \frac{a_1 cl}{b\{(2a_1 b/al)+3\}}\theta_1. \tag{7.9}$$

In the steady state the coordinate θ_1 is determined by the Lagrangian equation

$$\frac{\partial V}{\partial \theta_1} = Q, \tag{7.10}$$

where V is the thermal potential in the half-width of the web from A to M. Hence

$$V = \tfrac{1}{2}ca_1\int\limits_0^l \theta^2\,dy = \tfrac{1}{6}ca_1 l\theta_1^2. \tag{7.11}$$

The value of Q is determined by the principle of virtual work,

$$Q\,\delta\theta_1 = b\theta_0\,\delta H_b. \tag{7.12}$$

From equation (7.9) we derive δH_b in terms of $\delta\theta_1$. Hence

$$Q = \frac{a_1 cl}{(2a_1 b/al)+3}\theta_0. \tag{7.13}$$

With the values (7.11) and (7.13) for V and Q, equation (7.10) yields

$$\theta_1 = \frac{\theta_0}{(2a_1 b/3al)+1}. \tag{7.14}$$

Hence we have obtained the temperature θ_1 at the joint A in terms of the outside adiabatic temperature θ_0 at the flange.

The concept of effective flange-width and the corresponding associated field may be used to evaluate transient temperatures when θ_0 is time dependent, as outlined by the author.‡ It has also been verified that the method yields accurate solutions.

‡ See the author's paper cited on p. 81.

CHAPTER FIVE

NON-LINEAR SYSTEMS

1. INTRODUCTION

UNTIL now the analysis has been restricted to physical systems with properties independent of the temperature. In this chapter we shall consider cases where the heat capacity and thermal conductivity may depend on the temperature. Such systems are physically non-linear in the sense that the principle of superposition does not apply. Physical non-linearity may also be due to boundary conditions.

In order to generalize the variational principles to non-linear systems, Section 2 extends the concept of thermal potential to a medium with a temperature-dependent heat capacity. As shown in Section 3, this concept leads to a variational principle and Lagrangian equations for systems with temperature-dependent heat capacity and thermal conductivity.‡

The concept of associated field is extended to non-linear systems in Section 4. For the case where the thermal conductivity is constant and in the absence of a surface heat-transfer coefficient, the extension is immediate. It is also shown that the more general case, where the thermal conductivity is also a function of the temperature, may be reduced to one of constant thermal conductivity by a well-known transformation.

Non-linearity due to boundary conditions is discussed in Section 5. This will be the case for melting or freezing boundaries. An outline is given of the solution of the problem of ablation where a melted solid is removed at a melting boundary. It is also pointed out how non-linearity is introduced by surface radiation for large temperature variations.

In Section 6, as an example of a non-linear problem, we have evaluated numerically the penetration of heat in a medium with temperature-dependent properties. Because of non-linearity the cooling and heating problems are not the same and it is shown how they differ.

‡ These variational principles for non-linear systems were developed in the author's paper, 'New methods in heat flow analysis with application to flight structures', *J. aeronaut. Sci.* **24**, 857–73 (1957).

2. THERMAL POTENTIAL OF NON–LINEAR SYSTEMS

The concept of thermal potential of a given domain τ of a solid was discussed and used in the foregoing chapters in the context of linear systems. We shall consider a solid in which the heat capacity $c(x, y, z, \theta)$ per unit volume depends on the temperature θ and the coordinates x, y, z.

The concept of thermal potential may be extended to this case by introducing the *heat content* h per unit volume, defined as

$$\int_0^\theta c\, d\theta = h. \tag{2.1}$$

The heat content $h(x, y, z, \theta)$ is a function of the coordinates and the temperature, and represents a material property. Similarly, we define

$$\int_0^h \theta\, dh = F. \tag{2.2}$$

This quantity is also a function $F(x, y, z, \theta)$ of the coordinates and the temperature. It generalizes to a non-linear system the concept of *thermal potential* of a solid particle *per unit volume* for a linear system. When $c(x, y, z)$ is independent of the temperature the value of F becomes

$$F = \tfrac{1}{2}c\theta^2, \tag{2.3}$$

which coincides with the integrand in expression (2.8) of Chapter 1. The total thermal potential of a solid occupying a volume τ is defined as

$$V = \iiint_\tau F\, d\tau. \tag{2.4}$$

It should be pointed out that the volume dilatation is neglected.

3. VARIATIONAL PRINCIPLE

With the definition (2.4) of the thermal potential it is possible to formulate a variational principle in exactly the same form as in Chapter 1 for the case of a linear system.

Attention is called to the very general properties of the thermal conductivity assumed in the present case. It may be anisotropic, and a function not only of the coordinates x, y, z but also of both the temperature θ and the time t. Hence we write the thermal conductivity tensor as

$$k_{ij} = k_{ij}(x, y, z, t, \theta). \tag{3.1}$$

A time dependence of the thermal conductivity may have to be considered in special cases where the physical properties of the material are influenced by external factors, such as certain types of radiation.

In the formulation of the variational principle we shall use, as before, the inverse matrix of the thermal conductivity k_{ij} defined by equation (5.7) of Chapter 1. This inverse yields the thermal resistivity tensor which we write as

$$\lambda_{ij} = \lambda_{ij}(x, y, z, t, \theta). \tag{3.2}$$

The Onsager relations are assumed to be valid. Hence

$$\lambda_{ij} = \lambda_{ji}. \tag{3.3}$$

The variational formulation is obtained by introducing the heat displacement field with cartesian components,

$$H_i = H_i(x, y, z, t).$$

It is the same as defined by equation (5.1) in Chapter 1. We shall assume that it satisfies the equation

$$h = -\sum^{i} \frac{\partial H_i}{\partial x_i}. \tag{3.4}$$

This equation, which generalizes the conservation equation (5.2) of Chapter 1, may be considered as a holonomic constraint between the heat content h and the heat displacement H_i. With these definitions we will now show that the variational principle for the non-linear case is written

$$\delta V + \iiint_\tau \Big(\sum^{ij} \lambda_{ij} \dot{H}_j \delta H_i\Big)\, d\tau = -\iint_A \sum^{i} \theta n_i \,\delta H_i \, dA, \tag{3.5}$$

where V is given by equation (2.4) and the integrals are extended to a volume τ of the solid and its boundary A. From the definition (2.4) and equation (3.4) we derive

$$\delta V = \iiint_\tau \theta\, \delta h \, d\tau = -\iiint_\tau \theta \sum^{i} \frac{\partial}{\partial x_i} \delta H_i \, d\tau. \tag{3.6}$$

After integration by parts we obtain

$$\delta V = \iiint_\tau \sum^{i} \frac{\partial \theta}{\partial x_i} \delta H_i \, d\tau - \iint_A \sum^{i} \theta n_i \,\delta H_i \, dA. \tag{3.7}$$

Substitution into equation (3.5) yields

$$\iiint_\tau \sum^{i} \Big(\frac{\partial \theta}{\partial x_i} + \sum^{j} \lambda_{ij} \dot{H}_j\Big) \delta H_i \, d\tau = 0. \tag{3.8}$$

This equation is identical to the law of heat conduction,

$$\frac{\partial \theta}{\partial x_i} + \sum^{j} \lambda_{ij} \dot{H}_j = 0. \tag{3.9}$$

It is the same as equation (5.6) of Chapter 1 for the linear system. The only difference in the present case resides in the fact that λ_{ij} may be a function of the temperature.

Thus we have established the validity of the variational principle (3.5) for non-linear systems. Note that physically the principle is equivalent to stating that the rate of flow field $\dot{H}_i$ verifies the law of heat conduction under the given instantaneous distribution of temperature, while conservation of energy is verified identically by the formulation itself in analogy with holonomic constraints in mechanics. In the present case the holonomic constraint is expressed by equation (3.4).

The unknown field H_i may be defined in terms of generalized coordinates q_k as

$$H_i = H_i(q_1, q_2,..., q_n, x, y, z, t). \tag{3.10}$$

The variation principle (3.5) then leads to a set of n Lagrangian equations:

$$\frac{\partial V}{\partial q_i} + \frac{\partial D}{\partial \dot{q}_i} = Q_i. \tag{3.11}$$

Since the variational principle (3.5) is formally the same as in the linear case, the derivation of equations (3.11) follows exactly the same procedure as in Section 5 of Chapter 1. The dissipation function D and thermal force Q_i are defined by equations (5.18) and (5.22) of that chapter.

Principle of virtual work and minimum dissipation

As already pointed out in the first chapter, the variational principle may be interpreted as a principle of virtual work. This remark obviously applies to the variational expression (3.5) for non-linear systems. In particular, the generalized thermal forces are obtained by evaluating the variation $\sum^{i} Q_i \delta q_i$, which extends to thermodynamics the concept of virtual work.

The non-linear Lagrangian equations (3.11) are also equivalent to a principle of minimum dissipation as already shown in Section 4 of Chapter 1 for the linear case. For the non-linear system we may define, as before, a disequilibrium force

$$X_i = Q_i - \frac{\partial V}{\partial q_i}. \tag{3.11 a}$$

Equations (3.11) then express the condition that the dissipation function D is a minimum for all possible values of $\dot{q}_i$ such that $\sum^{i} X_i \dot{q}_i$ is a constant. It should be pointed out that the dissipation function is a positive-definite quadratic form in the rate variables. This is a consequence of the fact that D is proportional to the local entropy production.

Finite element method for non-linear systems

The numerical analysis of a transient field by dividing the domain into finite elements may be extended to non-linear systems. The procedure is similar to the one outlined in Section 7 of Chapter 3. The Lagrangian equations (3.11) are formulated for each sub-system. The thermal forces Q_i at the interconnecting boundaries may be grouped in pairs of opposite sign and equal absolute value. By equating these values after a change in sign, we derive a set of equations for the total system from which the unknowns Q_i at the interconnecting boundaries have been eliminated.

4. ASSOCIATED FIELDS FOR NON-LINEAR SYSTEMS

The concept of ignorable coordinates may be extended to non-linear systems. Consider a medium with temperature-dependent heat capacity, while the thermal conductivity tensor and the surface heat transfer coefficient are independent of the temperature. Let us write the heat displacement as in equation (2.1) of Chapter 4:

$$\mathbf{H} = \sum^{i} \mathbf{\Theta}_i q_i + \sum^{l} \mathbf{F}_l f_l. \tag{4.1}$$

The dissipation function in this case is a quadratic form with constant coefficients which may be written as

$$D = \tfrac{1}{2} \sum^{ij} b_{ij} \dot{q}_i \dot{q}_j + \sum^{il} b'_{il} \dot{q}_i \dot{f}_l + \tfrac{1}{2} \sum^{lm} b''_{lm} \dot{f}_l \dot{f}_m. \tag{4.2}$$

This coincides with expression (2.8) of Chapter 4 where, in the summations, i and j assume all values from 1 to ν, while l and m assume all values from 1 to k. As in Chapter 4, we assume that the k coordinates f_l do not appear in the thermal potential V. The Lagrangian equations are separated into two groups,

$$\frac{\partial V}{\partial q_i} + \frac{\partial D}{\partial \dot{q}_i} = Q_i, \qquad \frac{\partial D}{\partial \dot{f}_l} = Q_l. \tag{4.3}$$

The equations in the first group containing V are non-linear. This is because the heat capacity depends on the temperature. However, the coordinates q_i and f_l may be decoupled by exactly the same transformation as in the linear case. This is obvious if we note that the required transformation is given by equation (2.11) of Chapter 4. It involves only the dissipation function (4.2), which is the same in both the linear and non-linear cases. Hence f'_i are also ignorable coordinates for the non-linear problem with temperature-independent thermal conductivity.

As in the linear case the process of decoupling introduces the associated field as a vector field determined by a given temperature field.

The procedure of deriving the associated field by a principle of minimum dissipation is also applicable to this non-linear case and is exactly the same as in Section 3 of Chapter 4.

Finally, the analogue model for associated fields using a distribution of heat sinks as derived in Section 4 of Chapter 4 may be extended to the non-linear case. The only difference arises from the replacement of equation (4.6) of that chapter by the following

$$h = \operatorname{div}(k \operatorname{grad} \psi), \tag{4.4}$$

where h is the heat content (2.1). By following the same argument as in Chapter 4, it is shown that the field associated with a temperature distribution θ is the rate of flow field in a steady state with distributed sinks of magnitude

$$-w = h = \int_0^\theta c(\theta)\, d\theta \tag{4.5}$$

and boundary conditions as in the linear case.

Temperature-dependent conductivity

The method of associated fields may be extended to the case where the thermal conductivity depends on the temperature. This is accomplished by a well-known transformation of the heat diffusion equation. Consider first the case of isotropic conductivity. The temperature is governed by the equation

$$\operatorname{div}(k \operatorname{grad} \theta) = c\frac{\partial \theta}{\partial t}. \tag{4.6}$$

The thermal conductivity $k(\theta)$ and the heat capacity $c(\theta)$ per unit volume are assumed to be functions of the temperature. For simplicity, we write $c(\theta)$ instead of $c(x, y, z, \theta)$, and it is understood that the heat capacity may also be a function of the location. We introduce the variable,

$$u(\theta) = \int_0^\theta \frac{k(\theta)}{k_0}\, d\theta. \tag{4.7}$$

We choose k_0 to represent the thermal conductivity at a given reference temperature, for example $\theta = 0$. Hence

$$k_0 = k(0). \tag{4.8}$$

The transformation (4.7) therefore amounts to a change in the temperature scale. We derive the relations,

$$k_0 \operatorname{grad} u = k \operatorname{grad} \theta, \qquad \frac{\partial u}{\partial t} = \frac{k}{k_0}\frac{\partial \theta}{\partial t}. \tag{4.9}$$

Equation (4.6) becomes

$$k_0 \operatorname{div}(\operatorname{grad} u) = c\frac{k_0}{k}\frac{\partial u}{\partial t}. \tag{4.10}$$

This equation governs the temperature u in a medium of constant thermal conductivity k_0 and heat capacity $c'(u)$ a function of u given by

$$c'(u) = \frac{k_0}{k(\theta)}c(\theta). \tag{4.11}$$

As pointed out at the beginning of this section, the method of associated fields is applicable for constant conductivity k_0. Hence it may be used if the fictitious medium represented by equation (4.10) is analysed instead of the actual physical system. The temperature u in the fictitious medium is different from θ, but the heat displacement $\mathbf{H}$ and heat content h are the same as the actual values. The heat content is

$$h = \int_0^u c'(u)\,du = \int_0^\theta c(\theta)\,d\theta. \tag{4.12}$$

Note that a straightforward application of the method of associated fields in the fictitious medium requires that the thermal system does not include a surface heat-transfer coefficient. This can be seen by going back to the definition of the surface heat-transfer coefficient K. According to equation (2.1) of Chapter 2, we write

$$\dot{H}_n = K(\theta - \theta_a). \tag{4.13}$$

This relation may also be written

$$\dot{H}_n = K'(u - u_a), \tag{4.14}$$

where $$u = u(\theta), \qquad u_a = u(\theta_a), \qquad K'(u) = K\frac{\theta-\theta_a}{u-u_0}. \tag{4.15}$$

Hence the use of the fictitious temperature u requires the introduction of a temperature-dependent surface heat-transfer coefficient $K'(u)$. Therefore, when the thermal conductivity depends on the temperature, the associated field should be evaluated for the solid as a separate system excluding the surface heat-transfer. The latter may then be included in a separate evaluation of the generalized forces applied directly to the solid itself.

The method of associated fields may also be extended to a system where the thermal conductivity is both anisotropic and temperature-dependent, provided the conductivity tensor is of the form

$$k_{ij}(\theta) = k'_{ij} f(\theta), \tag{4.16}$$

where k'_{ij} are constants. If we assume that $f(\theta)$ is a non-dimensional function of temperature, we derive an analogue model of thermal conductivity k'_{ij} and a fictitious temperature

$$u = \int_0^\theta f(\theta)\, d\theta. \tag{4.17}$$

The derivation follows exactly the same procedure as for equation (4.6).

Approximate method of associated fields

In some problems, where the thermal conductivity depends on the temperature, an approximate method of decoupling the ignorable coordinates may be suggested. Instead of using the transformation (4.7) we assume the thermal conductivity to be equal to some average temperature-independent value. This average value may depend on the location. The associated field is then evaluated assuming such average values. However, in evaluating the dissipation function in terms of these associated fields the actual temperature-dependent conductivity is used. This approximate method, in contrast to that based on the transformation (4.7), does not require that the thermal conductivity be independent of the location, and includes the surface heat-transfer properties in the analysis.

5. MELTING BOUNDARIES AND RADIATION

Non-linear features may appear due to other reasons than temperature dependence of material properties. Non-linearity may be introduced by boundary conditions. This will be the case, for example, in problems with melting or freezing boundaries. The motion of the boundary depends on the unknown transient temperature field. The position of the boundary must therefore be introduced into the equations as an additional unknown. Radiation at a boundary constitutes another example of non-linearity due to boundary conditions.

Melting boundaries

We shall outline briefly the formulation of a problem with melting boundaries by treating a simple example.

As already pointed out with reference to linear systems, the variational principle may be applied by assuming that in the definition of the thermal potential and the dissipation function the volume of integration has moving boundaries. The corresponding thermal force is then defined on a corresponding moving surface. This property is a consequence of the fact that the Lagrangian equations govern the thermal flow for given

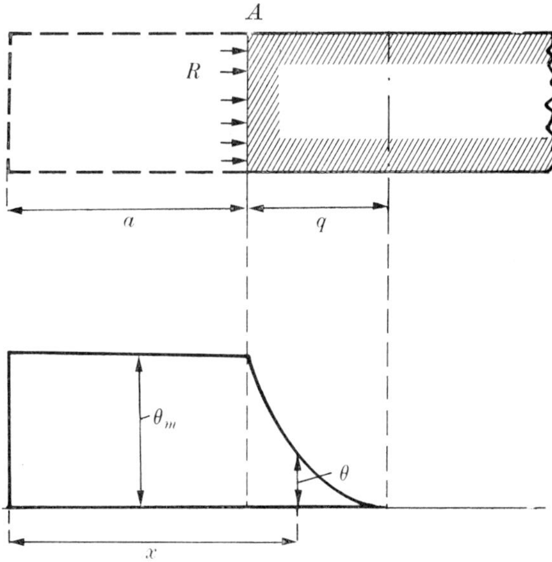

FIG. 5.1. Ablation of half-space subject to constant rate of heat input R at the melting surface.

instantaneous configurations of boundaries and temperature irrespective of the time history. For the same reason, the principle is also applicable for moving boundaries in systems with temperature-dependent properties. The foregoing remarks indicate that the variational principle and the Lagrangian equations may be used to solve problems of this type.

As an illustration of the method, consider the simpler problem of ablation for a medium with temperature-independent properties. In problems of ablation the melted material is removed from the boundary as soon as the melting occurs. A solid half-space of constant thermal conductivity k and constant heat capacity c is melting at the surface A (Fig. 5.1). Heat is injected at the melting surface at a rate R per unit time and unit area. At the time t the material has been removed by melting to a depth $a(t)$. This is a case of moving boundaries and, as pointed out, the Lagrangian equations are applicable to this case.

The transient temperature distribution in the solid is approximated by the cubic

$$\theta = \theta_m\left\{1-\frac{(x-a)}{q}\right\}^3. \tag{5.1}$$

There are two unknowns here, the depth of melting $a(t)$ and the penetration depth $q(t)$ of the heat into the unmelted solid. The coordinate along the depth is x and the melting temperature is $\theta = \theta_m$ at $x = a$.

An important point here, in applying the Lagrangian equation, is that only q is considered as a generalized coordinate, while $a(t)$ is treated as if it were a given function of time not subject to variations when applying the variational principle. In fact, the depth of melting $a(t)$ is also an unknown. However, it will have to be determined by an *auxiliary equation* that is not part of the variational process.

The heat displacement at the depth x is

$$H = \int\limits_x^{a+q} c\theta\, dx = \tfrac{1}{4}qc\theta_m\left\{1-\frac{(x-a)}{q}\right\}^4. \tag{5.2}$$

When evaluating the thermal potential, the dissipation function, and the thermal force, we integrate over the region $a < x < a+q$. We must evaluate

$$V = \tfrac{1}{2}\int\limits_a^{a+q} c\theta^2\, dx, \qquad D = \frac{1}{2}\int\limits_a^{a+q}\frac{1}{k}\dot{H}^2\, dx, \qquad Q = \theta_m\left(\frac{\partial H}{\partial q}\right)_{x=a}, \tag{5.3}$$

and derive the corresponding Lagrangian equation

$$\frac{\partial V}{\partial q}+\frac{\partial D}{\partial \dot{q}} = Q. \tag{5.4}$$

By introducing the values (5.3), equation (5.4) is written

$$\left(\frac{4}{112}\dot{q}+\frac{11}{112}\dot{a}\right)q = \frac{5}{14}\frac{k}{c}. \tag{5.5}$$

It contains two unknowns, a and q. An auxiliary equation is obtained from the condition of conservation of energy in the melting process. It is written

$$R = (L+c\theta_m)\dot{a}+\tfrac{1}{4}\theta_m c\dot{q}, \tag{5.6}$$

where L is the latent heat of melting per unit volume. Hence we have obtained two simultaneous differential equations, (5.5) and (5.6), for the two unknown functions of time, $q(t)$ and $a(t)$. Equations (5.5) and (5.6) were derived and solved numerically in an earlier paper.‡ In later

‡ M. A. Biot and H. Daughaday, 'Variational analysis of ablation', *J. Aerospace Sci.* **29**, 228–9 (1962).

treatment‡ the variational method was applied to a similar ablation problem for large temperature variations and a strong dependence of the thermal conductivity on the temperature.

The melting of a semi-infinite solid initially at the melting temperature without removal of the liquid phase has been solved using the same variational method in a paper by Lardner.§

Surface radiation

Physical non-linearity will also occur when surface radiation takes place with large temperature variations. In such cases the surface heat-transfer properties may not be linearized. The rate of heat flow per unit area due to surface radiation may be written

$$\dot{H}_r = \epsilon\sigma\{(T_e+\theta)^4 - T_e^4\}, \tag{5.7}$$

where ϵ is the emissivity, σ the Stefan constant, and T_e is the absolute equilibrium temperature for which no radiation loss occurs. The non-linear boundary condition is then

$$\dot{H}_r = \dot{\mathbf{H}}\,.\,\mathbf{n}, \tag{5.8}$$

where $\mathbf{n}$ is the unit normal at the boundary and $\mathbf{H}$ is the heat displacement in the solid at the boundary.

For sufficiently small values of the excess temperature equation (5.7) may be linearized and becomes

$$\dot{H}_r = K\theta, \tag{5.9}$$

where

$$K = 4\epsilon\sigma T_e^3 \tag{5.10}$$

plays the role of a temperature-independent surface heat-transfer coefficient as defined in Section 2 of Chapter 2.

In some problems dealing with non-linear radiation it is sometimes advantageous to apply the boundary condition to the temperature field. For example, for isotropic conductivity k the approximate temperature field is chosen so as to satisfy the boundary condition

$$k\,\text{grad}_n\,\theta + \epsilon\sigma\{(T_e+\theta)^4 - T_e^4\} = 0. \tag{5.11}$$

This procedure is illustrated in a paper by Rafalski and Zyskowski.|| They propose to subject equation (5.11) to some averaging process over the boundary, thus imposing a holonomic constraint on the unknowns q_i.

‡ M. A. Biot and H. C. Agrawal, 'Variational analysis of ablation for variable properties', *J. Heat Transfer*, **86**, 437–42 (1964).

§ T. J. Lardner, 'Approximate solutions to phase-change problems', *AIAA Jl* **5**, 2079–80 (1967).

|| P. Rafalski and W. Zyskowski, 'Lagrangian approach to the non-linear boundary heat-transfer problem', *AIAA Jl* **6**, 1606–8 (1968).

6. HEATING AND COOLING OF A WALL WITH NON-LINEAR PROPERTIES

As an illustration we shall treat a very simple non-linear problem. Consider a semi-infinite body that occupies the region $x > 0$. The surface $x = 0$ is brought suddenly to a constant temperature $\theta = \theta_0$

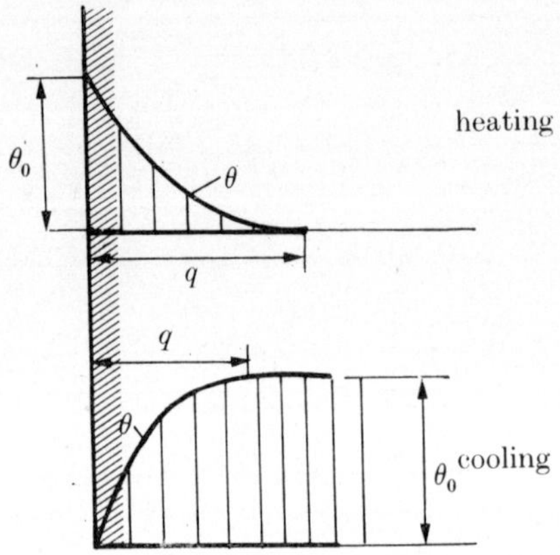

FIG. 5.2. Heating and cooling of a wall with non-linear properties.

at the time $t = 0$ (Fig. 5.2). We assume the heat capacity to be a linear function of the temperature as follows:

$$c(\theta) = c_0\left(1+\frac{\theta}{\theta_0}\right). \tag{6.1}$$

The thermal conductivity k is assumed constant.‡ However, this assumption does not restrict the generality of the example since, as shown in Section 4, the case of temperature-dependent conductivity is easily transformed to one with a constant value. The temperature distribution in the solid is approximated by the expression

$$\theta = \theta_0\left(1-\frac{x}{q}\right)^2, \tag{6.2}$$

where the penetration depth q plays the role of a generalized coordinate. The problem is analysed by considering a cylindrical volume of unit cross-section parallel to the x axis. The heat content (2.1) is

$$h = \int\limits_0^\theta c\,d\theta = c_0\,\theta+\tfrac{1}{2}c_0\frac{\theta^2}{\theta_0}. \tag{6.3}$$

‡ We shall follow the treatment of this problem given in the author's paper, 'New methods in heat flow analysis with application to flight structures', *J. aeronaut. Sci.* **24,** 857–73 (1957).

The value of F and the thermal potential (2.4) are

$$F = \int_0^\theta c\theta\, d\theta = \tfrac{1}{2}c_0\,\theta^2 + \tfrac{1}{3}c_0\frac{\theta^3}{\theta_0}, \qquad V = \int_0^q F\, dx = \tfrac{31}{210}c_0\,\theta_0^2\, q. \tag{6.4}$$

The heat displacement is obtained from the relation

$$H = \int_x^q h\, dx. \tag{6.5}$$

By putting

$$\zeta = 1 - \frac{x}{q} \tag{6.6}$$

we derive

$$H = (\tfrac{1}{3}\zeta^3 + \tfrac{1}{10}\zeta^5)c_0\,\theta_0\, q; \tag{6.7}$$

the dissipation function is

$$D = \frac{1}{2k}\int_0^q \dot{H}^2\, dx = \frac{0{\cdot}0648}{2k}\, c_0^2\,\theta_0^2\, q\dot{q}^2. \tag{6.8}$$

The virtual work of the thermal force is

$$Q\,\delta q = \theta_0\,\delta H, \tag{6.9}$$

where δH is the variation of H at $x = 0$. Hence

$$Q = \tfrac{13}{30}c_0\,\theta_0^2. \tag{6.10}$$

The Lagrangian equation is

$$\frac{\partial V}{\partial q} + \frac{\partial D}{\partial \dot{q}} = Q. \tag{6.11}$$

With the values (6.4), (6.8), and (6.10), this equation becomes

$$0{\cdot}0648 q\dot{q} = \frac{2}{7}\frac{k}{c_0}. \tag{6.12}$$

Integration of this equation, with the initial condition $q = 0$ for $t = 0$, yields

$$q = 2{\cdot}97\sqrt{\left(\frac{kt}{c_0}\right)}. \tag{6.13}$$

We may compare this result with the value

$$q = 3{\cdot}36\sqrt{\left(\frac{kt}{c}\right)} \tag{6.14}$$

given by equation (7.10) of Chapter 1 for the linear case when k and c are both constant. The values (6.13) and (6.14) can be made to coincide by putting

$$c = 1{\cdot}28c_0. \tag{6.15}$$

Hence the non-linear case is approximated by assuming that c is constant and equal to the value (6.15). Note that this value lies about half-way between the average value $1{\cdot}5c_0$ and the value c_0 at the lowest temperature.

The problem of cooling from $\theta = \theta_0$ to $\theta = 0$ of the half-space of the same non-linear material (Fig. 5.2) was also analysed in the paper already quoted.‡ Problems of heating and cooling are not the same in non-linear problems. In this case it is found that the penetration depth is the same as in a medium of constant value of c equal to $c = 1{\cdot}74c_0$. Hence in this case the constant equivalent value of c lies midway between the average value $1{\cdot}5c_0$ and the value $2c_0$ at the highest temperature.

‡ See p. 96.

CHAPTER SIX

CONVECTIVE HEAT TRANSFER

1. INTRODUCTION

In this chapter the variational principles and Lagrangian equations are extended to include heat transfer by convection. We shall consider two different approaches. In the first we treat the problem of heat conduction in a solid whose boundaries are in contact with a moving fluid. The convective heat transfer at the boundary is taken into account by means of a 'trailing function'. This trailing function embodies the convective properties of the fluid, which may then be incorporated in the boundary condition. By this procedure the convective part of the problem is separated from conduction in the solid. The flow may be laminar or turbulent. An evaluation of the trailing function is carried out in Chapter 7. Some advantage results from this separation because the physical properties of conduction and convection are quite different and, in general, are not conveniently treated by the same analysis.

The second approach is to derive unified equations applicable to complex systems that are composed partly of solid and partly of moving fluids. This unified approach should be useful in a number of specialized cases, such as those of heat transfer in a porous solid through which a fluid is in motion.

The trailing function is described in Section 2. It represents, essentially, the temperature perturbation at a fluid–solid boundary due to heat injection into the moving fluid at a given point of the interface. It may be defined as a two-dimensional or a one-dimensional concept. In Section 3 additional terms are derived for the Lagrangian equations to include the convective heat transfer at the boundary in terms of trailing functions.

The method of associated fields discussed in Chapter 4 is extended in Section 4 to include convective heat transfer. The concept of ignorable coordinates is applicable to this case and leads to a considerable reduction of the number of generalized coordinates required in the analysis. A method analogous to the principle of minimum dissipation is also applicable for the evaluation of the associated field, and is shown to be related to the properties of an adjoint system obtained by reversing the flow.

The second fundamental approach is derived in Section 5. By a suitable definition of the dissipation function it is shown that the Lagrangian equations are valid for both convection and conduction. Thus a single form of equations governs mixed systems including solids and moving fluids with laminar or turbulent flow. The principle of minimum dissipation is therefore applicable to convective heat transfer.

2. TRAILING FUNCTION

Consider a solid boundary in contact with a moving fluid. In order to describe the general properties of heat transfer between the solid and the moving fluid the author has introduced the concept of *trailing function*‡ by a very general definition. The concept fits quite naturally into an extension of the Lagrangian equations for heat conduction in a solid to include the case of convective heat transfer at the boundary when the concept of local heat-transfer coefficient breaks down. That convective heat transfer cannot adequately be described in terms of local heat-transfer coefficients was demonstrated by the author in a detailed physical discussion.§

The flow may be laminar or turbulent. In the case of turbulent flow the velocity and temperature field are defined by averaging the fluctuations. The velocity field is treated as being a given function of time and of the coordinates, while the temperature is an unknown transient or steady field to be determined as part of the heat transfer problem.

In order to formulate the coupled problem of heat transfer between the solid and the fluid, it is necessary to introduce the so-called adiabatic temperature θ_a already discussed in a more restricted case in Section 2 of Chapter 2. It is defined as the temperature distribution of the fluid at the solid boundary when no heat flow occurs across the interface. In other words, we consider the temperature field that would arise in the fluid if the solid boundary were thermally insulated. This temperature field may be a given function of time and will depend on certain external or internal factors that will cause heating or cooling of the fluid and may be the result of radiation, internal friction, chemical reactions, up-stream heating, etc. Under these conditions the adiabatic temperature $\theta_a(P, t)$ is the temperature of the fluid at the solid boundary, considered as a function of the time t and the location P at the surface of the solid.

‡ M. A. Biot, 'Lagrangian thermodynamics of heat transfer in systems including fluid motion,' *J. Aerospace Sci.* **29,** 568–77 (1962).

§ M. A. Biot, 'Fundamentals of boundary-layer heat transfer with streamwise temperature variations', ibid. 558–67 (1962).

In order to derive a formulation of the heat transfer for the case where the boundary is not thermally insulated, let us first assume steady-state conditions, hence time-independent flow and temperatures. In this case heat flow occurs across the interface, and the temperature of the fluid at the surface is now θ instead of the adiabatic temperature θ_a. The basic assumption is introduced that the heat flow through the interface is linearly related to the temperature change $\theta-\theta_a$ at the surface by the integral relation

$$\theta-\theta_a = \iint_A \dot{H}_n(P')r(P,P')\,dA_{P'}. \tag{2.1}$$

In this expression θ and θ_a are respectively the actual and adiabatic temperatures at a given point P of the boundary. The surface integral is extended to all surface elements $dA_{P'}$ at points P' of the boundary, while $\dot{H}_n(P')$ is the rate of heat flow per unit time and unit area from the solid into the fluid at the point P'. While relation (2.1) is linear it does not imply that thermal properties of the fluid are governed by linear laws. It is only required that the thermal perturbation due to the surface heat transfer satisfies the principle of superposition within a certain temperature range.

The surface heat-transfer properties are embodied in the two-point function $r(P,P')$. In this general context it was introduced by the author under the designation of *trailing function.*

The reason for this designation and the physical significance of this function are brought out by considering a concentrated heat injection equal to unity per unit time a point P'. The temperature increment at a point P of the boundary is then

$$\theta-\theta_a = r(P,P'). \tag{2.2}$$

For a given point P' as origin, this expression represents a two-dimensional temperature field. In general, the isothermal contours of this temperature field will be similar to the pattern shown in Fig. 6.1, corresponding to a 'trail' down-stream of the heat source.

Surface heat-transfer was analysed in Section 2 of Chapter 2 for the particular case where the concept of the local heat-transfer coefficient K is applicable. This may be considered as a limiting case of the more general concept of trailing function by using a Dirac function $\delta(P,P')$ with the properties

$$\begin{aligned} &\delta(P,P') = 0 \quad \text{for } P \neq P', \\ &\iint_A \delta(P,P')\,dA_{P'} = 1. \end{aligned} \tag{2.3}$$

The surface heat-transfer coefficient $K(P)$ is equivalent to the trailing function

$$r = \frac{\delta(P, P')}{K(P)}. \tag{2.4}$$

This can be verified by substituting this value of r into expression (2.1). We obtain

$$(\theta - \theta_a)K(P) = \dot{H}_n(P). \tag{2.5}$$

This result coincides with equation (2.1) of Chapter 2.

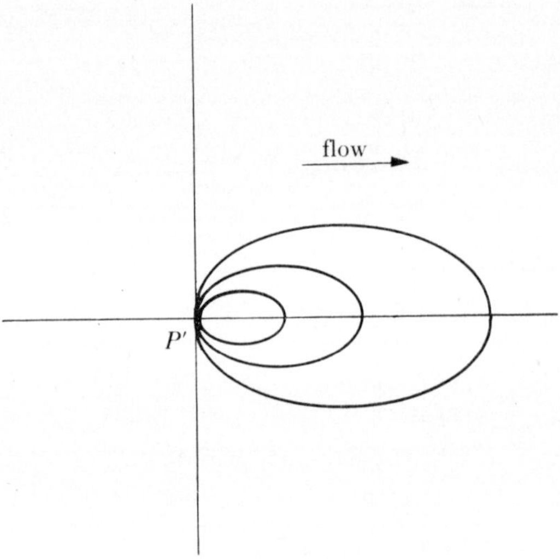

FIG. 6.1. Isothermal contour lines for the trailing function $r(P, P')$ with heat injection at the point P' (shown schematically).

In order to simplify the presentation, we have assumed both the flow and the temperature field to be stationary. Let us now retain stationary flow condition and consider time-dependent temperatures. This means that the rate of heat injection $\dot{H}_n(P', t)$ may vary with time. Strictly speaking, the trailing function for such a case should involve a time lag, because of the finite velocity of down-stream convection. However, as shown in an earlier paper,‡ there is a large category of technological problems for which the time lag is negligible. In these cases the trailing function $r(P, P')$ is the same as for the steady state and equation (2.1) remains applicable. It becomes

$$\theta - \theta_a = \iint_A \dot{H}_n(P', t) r(P, P')\, dA_{P'}. \tag{2.6}$$

‡ See p. 100, n. §.

The next step is to assume the fluid motion to be a given function of time. In that case the adiabatic temperature $\theta_a(P,t)$ is a given function of time. The trailing function $r(P,P',t)$ also becomes a function of time and equation (2.2) is replaced by

$$\theta-\theta_a = r(P,P',t). \tag{2.7}$$

This expression represents the temperature increment at point P due to a constant concentrated unit rate of heat injection at point P'. It is assumed here that we may neglect the time lag. In that case the trailing function corresponds to quasi-steady conditions, and is the same as if the fluid motion were stationary and equal to the instantaneous velocity field at any particular instant t. Relation (2.6) becomes

$$\theta-\theta_a = \iint_A \dot{H}_n(P',t)r(P,P',t)\,dA_{P'}. \tag{2.8}$$

In a large category of problems the assumption of quasi-steady conditions for convective heat transfer will be valid.

In cases where the convective time lag must be taken into account a straightforward generalization of the trailing function is written as $r(P,P',t,t')$. It is defined as the temperature increment at the point P and time t due to a unit injection of heat at point P' and time t'. In the analysis that follows we will assume that the convective time lag is negligible and that the definition (2.7) of the trailing function is adequate.

One-dimensional trailing function

The concept of trailing function is considerably simplified if we assume a two-dimensional flow field. The solid boundary in this case is a cylindrical surface, the flow field being the same in all planes perpendicular to this surface. We denote by s the length of arc along the streamline on the surface, and by y the distance along the straight lines normal to the flow. The quantities s and y may be chosen as coordinates of a point on the surface. Heat is then injected into the fluid along a straight line normal to the flow and located at the coordinate s'. The rate of heat injection concentrated on this line is equal to unity per unit time and unit length measured along y. For steady flow the temperature change $\theta-\theta_a$ at a point s along the streamline is then

$$\theta-\theta_a = r(s,s'). \tag{2.9}$$

This is a one-dimensional trailing function. Note that its *physical dimensions* are different from those of the corresponding two-dimensional trailing function (2.2). This is because expression (2.9) is the temperature change due to a heat injection per unit length and concentrated on a line instead of at a point.

If the flow varies with time we may assume that the convection is equivalent to a sequence of instantaneous steady states. The one-dimensional trailing function is then

$$\theta - \theta_a = r(s, s', t). \tag{2.10}$$

Consider now a one-dimensional distribution of heat injection. We denote by $\dot{H}_n(s', t)$ the rate of heat injection into the fluid per unit area and unit time at the point s' and the time t. We assume that this value is the same at all points along straight lines normal to the flow. Hence it is independent of y. The temperature change of the fluid at the boundary along a streamline is then expressed as

$$\theta - \theta_a = \int^{s'} \dot{H}_n(s', t) r(s, s', t)\, ds'. \tag{2.11}$$

The integration is a line integral along s'. It is the one-dimensional equivalent of the surface integral of equation (2.8).

In many problems, where the flow is not exactly two-dimensional, the concept of one-dimensional trailing functions remains applicable as an approximation. We must assume in this case that the temperature does not vary too rapidly in a direction normal to the flow. We may then use expression (2.11) for the temperature change with line integrations evaluated along the streamlines.

3. LAGRANGIAN EQUATIONS FOR CONDUCTION WITH BOUNDARY CONVECTION

Consider the Lagrangian equations for conduction in the solid,

$$\frac{\partial V}{\partial q_i} + \frac{\partial D}{\partial \dot{q}_i} = Q'_i, \tag{3.1}$$

where Q'_i is the thermal force corresponding to the temperature θ at the solid boundary

$$Q'_i = -\iint_A \theta \frac{\partial H_n}{\partial q_i}\, dA. \tag{3.2}$$

In these equations the field components H_i are expressed as

$$H_i = H_i(q_1, q_2, \dots, q_n, x, y, z, t) \tag{3.3}$$

and the outward normal component of this field at the boundary is

$$H_n = H_n(q_1, q_2, \dots, q_n, x, y, z, t). \tag{3.4}$$

If we make the usual assumption that the temperature in the solid and the fluid coincide at the boundary we may substitute the value θ of equation (2.8) into the thermal force (3.2). We find

$$Q' = Q_i - C_i, \tag{3.5}$$

where

$$\begin{aligned} Q_i &= -\int_A \theta_a \frac{\partial H_n}{\partial q_i}\, dA, \\ C_i &= \int_A \int_A \frac{\partial H_n}{\partial q_i}(P)\, \dot{H}_n(P') r(P, P', t)\, dA_P\, dA_{P'}. \end{aligned} \tag{3.6}$$

For simplicity we have represented the surface integral extended to the boundary A by a single integral sign. For the same reason we have written $H_n(P)$, or simply H_n, instead of $H_n(P, t)$. The Lagrangian equations (3.1) may be written as

$$\frac{\partial V}{\partial q_i} + \frac{\partial D}{\partial \dot{q}_i} + C_i = Q_i. \tag{3.7}$$

The term Q_i plays the role of a thermal force which now corresponds to the adiabatic boundary temperature θ_a. The new term C_i embodies the surface heat-transfer characteristics of the moving fluid.

In order to bring out the significance of the term C_i, consider a linear problem with temperature-independent heat capacity and conductivity, and assume that the vector $\mathbf{H}$ is expressed by the linear representation (3.1) of Chapter 2 as

$$\mathbf{H} = \sum^{i} \mathbf{H}^{(i)}(x, y, z) q_i. \tag{3.8}$$

We derive

$$H_n(P) = \sum^{i} H_n^{(i)}(P) q_i, \qquad \dot{H}_n(P) = \sum^{i} H_n^{(i)}(P) \dot{q}_i. \tag{3.9}$$

In that case the value (3.6) of C_i becomes

$$C_i = \sum^{j} c_{ij} \dot{q}_j \tag{3.10}$$

with

$$c_{ij} = \int_A \int_A H_n^{(i)}(P) H_n^{(j)}(P') r(P, P', t)\, dA_P\, dA_{P'}. \tag{3.11}$$

The thermal potential and dissipation function in this linear case are

$$V = \tfrac{1}{2} \sum^{ij} a_{ij} q_i q_j, \qquad D = \tfrac{1}{2} \sum^{ij} b_{ij} \dot{q}_i \dot{q}_j. \tag{3.12}$$

Hence the Lagrangian equations (3.1) are written as

$$\sum^{j} a_{ij} q_j + \sum^{j} (b_{ij} + c_{ij}) \dot{q}_j = Q_i. \tag{3.13}$$

They differ from the linear equations (3.14) of Chapter 2 by the presence of the coefficients c_{ij}. These coefficients must be distinguished from the coefficients b_{ij} by the fact that in general they are not symmetric:

$$c_{ij} \neq c_{ji}. \tag{3.14}$$

As can be seen from the definition (3.11), this inequality is a consequence of the following fundamental property of the trailing function:

$$r(P, P', t) \neq r(P', P, t). \tag{3.15}$$

It is essentially due to the fluid motion. Reciprocity properties are therefore not valid in heat conduction problems that involve fluid motion at the boundary.

If the fluid is at rest, the inequality (3.15) is replaced by an equality and reciprocity relations are verified.

The case discussed in Chapter 2 (Section 2), where the surface heat-transfer is represented by a local heat-transfer coefficient K, may be considered as a particular case of this more general treatment. It is obtained by writing the trailing function in the form (2.4) as

$$r(P, P', t) = \frac{1}{K(P, t)} \delta(P, P'). \tag{3.16}$$

In this case $\delta(P, P') = \delta(P', P) = 0$ for $P \neq P'$. Hence the coefficients

$$c_{ij} = c_{ji} \tag{3.17}$$

are symmetric and may be incorporated in the dissipation function as already shown in Chapter 2.

4. ASSOCIATED FIELDS FOR CONVECTIVE HEAT TRANSFER

It was shown by the author‡ that the method of associated fields as developed in Chapter 4 may be extended to include convective heat transfer.

We shall consider the case where the thermal conductivity is isotropic and independent of the temperature. We write the heat displacement field as in equation (4.4) of Chapter 4.

$$\mathbf{H} = \mathbf{\Theta} + \mathbf{F}, \tag{4.1}$$

where

$$\mathbf{\Theta} = -k \operatorname{grad} \psi,$$

$$\psi = \psi(q_1, q_2, \ldots, q_\nu, x, y, z, t), \tag{4.2}$$

$$\mathbf{F} = \mathbf{F}(f_1, f_2, \ldots, f_k, x, y, z, t),$$

with

$$\operatorname{div} \mathbf{F} = 0. \tag{4.3}$$

‡ See p. 100, n. ‡.

The generalized coordinates are the ν coordinates q_i, and the k coordinates f_l. The problem is to decouple these two groups of coordinates in the Lagrangian equations. The dissipation function is written as

$$D = D_q + D_{qf} + D_f, \tag{4.4}$$

where

$$D_{qf} = -\iiint_\tau \dot{\mathbf{F}}\,.\,\mathrm{grad}\,\dot{\psi}\,d\tau \tag{4.5}$$

contains the coupling terms. The integral is extended to the volume τ. Integration by parts of expression (4.5), taking into account equation (4.3), yields

$$D_{qf} = -\int_A \dot{\psi}\dot{F}_n\,dA. \tag{4.6}$$

This expression represents a surface integral extended to the solid boundary A. Again here, for simplicity, we use a single integral sign instead of a double integral. We denote by F_n the outward normal component of $\mathbf{F}$ at the boundary A.

The Lagrangian equations (3.7) are separated into two groups:

$$\frac{\partial V}{\partial q_i} + \frac{\partial D}{\partial \dot{q}_i} + C_i = Q_i, \qquad \frac{\partial D}{\partial \dot{f}_l} + C_l = Q_l. \tag{4.7}$$

In these equations Q_i and Q_l are the thermal forces associated respectively with the coordinates q_i and f_l and due to the adiabatic temperature θ_a at the boundary. According to equations (3.6) the values of C_i and C_l are as follows:

$$\begin{aligned} C_i &= \int_A\int_A \frac{\partial \Theta_n}{\partial q_i}(P,t)\dot{H}_n(P',t)r(P,P',t)\,dA_P\,dA_{P'}, \\ C_l &= \int_A\int_A \frac{\partial F_n}{\partial f_l}(P,t)\dot{H}_n(P',t)r(P,P',t)\,dA_P\,dA_{P'}. \end{aligned} \tag{4.8}$$

In the first group of equations (4.7) the coupling terms are

$$\mathscr{C}_i = \frac{\partial D_{qf}}{\partial \dot{q}_i} + \int_A\int_A \frac{\partial \Theta_n}{\partial q_i}(P,t)\dot{F}_n(P',t)r(P,P',t)\,dA_P\,dA_{P'}. \tag{4.9}$$

By interchanging P and P' in expression (4.9) we obtain

$$\mathscr{C}_i = \frac{\partial D_{qf}}{\partial \dot{q}_i} + \int_A\int_A \frac{\partial \Theta_n}{\partial q_i}(P',t)\dot{F}_n(P,t)r(P',P,t)\,dA_P\,dA_{P'}. \tag{4.10}$$

On the other hand,

$$\dot{\psi} = \sum^i \frac{\partial \psi}{\partial q_i}\dot{q}_i + \frac{\partial \psi}{\partial t}; \tag{4.11}$$

hence
$$\frac{\partial D_{qf}}{\partial \dot{q}_i} = -\int_A \frac{\partial \psi}{\partial q_i}(P,t)\dot{F}_n(P,t)\,dA_P. \tag{4.12}$$

With this value, expression (4.10) for $\mathscr{C}_i$ becomes
$$\mathscr{C}_i = \int_A \frac{\partial \mathscr{L}}{\partial q_i}\dot{F}_n(P,t)\,dA_P, \tag{4.13}$$

where
$$\mathscr{L} = -\psi(P,t)+\int_A \Theta_n(P',t)r(P',P,t)\,dA_{P'}. \tag{4.14}$$

The coordinates q_i and f_l are decoupled in expression (4.9) if we put $\mathscr{L} = 0$, i.e. if we choose ψ in such a way that it satisfies the boundary condition
$$\psi(P,t) = \int_A \Theta_n(P',t)r(P',P,t)\,dA_{P'}. \tag{4.15}$$

It is readily verified that under the same condition the coordinates q_i and f_l are also decoupled in the second group of equations (4.7), provided the trailing function r is independent of t.

Associated fields and analogue model

Consider a temperature field
$$\theta = \theta(q_1, q_2,..., q_\nu, x, y, z, t) \tag{4.16}$$
function of ν generalized coordinates. It is related to the heat displacement field (4.1) by the relation
$$c\theta = -\operatorname{div}\mathbf{H}. \tag{4.17}$$
We define a heat displacement field $\mathbf{\Theta}'$ associated with θ by putting
$$\mathbf{H} = \mathbf{\Theta}'+\mathbf{F}, \tag{4.18}$$
where
$$\mathbf{\Theta}' = -k\operatorname{grad}\psi, \qquad \operatorname{div}\mathbf{F} = 0, \tag{4.19}$$
and by assuming that ψ satisfies the boundary condition (4.15). From these relations we derive
$$c\theta = \operatorname{div}(\operatorname{grad}\psi). \tag{4.20}$$
When the field θ is given, this equation, along with the boundary condition
$$\psi(P,t) = \int_A \Theta'_n(P',t)r(P',P,t)\,dA_{P'}, \tag{4.21}$$
completely defines ψ. Hence the associated field $\mathbf{\Theta}'$ is determined by the temperature θ. By using this associated field the ν coordinates q_i are decoupled from the ignorable coordinates f_l in the first of equations (4.7).

The boundary condition (4.21) may be interpreted physically by writing

$$\bar{r}(P, P', t) = r(P', P, t). \tag{4.22}$$

Equation (4.21) then becomes

$$\psi(P,t) = \int_A \Theta'_n(P',t)\bar{r}(P,P',t)\,dA_{P'}. \tag{4.23}$$

Comparison with equation (2.8) shows that $\psi(P,t)$ is the rise of temperature at the boundary in the presence of a moving fluid with a trailing function $\bar{r}(P, P', t)$. In addition, equation (4.20) governs the temperature ψ in a volume with distributed and time-independent heat sinks of magnitude $c\theta$ per unit volume. Hence we obtain for the associated field an analogue model similar to that obtained in Section 4 of Chapter 4. The difference lies in the boundary condition. The field $\mathbf{\Theta}'$ is the rate of flow due to steady-state heat sinks and a boundary condition given by the trailing function (4.22). It should be noted that this analogue model represents an instantaneous steady state condition frozen for a particular instant t.

Further physical interpretation of the analogue model is derived by noting that in many cases, as shown below, the function $\bar{r}(P, P', t)$ appearing in the boundary condition (4.23) is the trailing function corresponding to the case of convective heat transfer obtained by reversing the fluid velocity field.

Variational principle for associated fields

For the non-convective heat transfer the associated field $\mathbf{\Theta}'$ may be derived from the temperature θ by minimizing the dissipation as shown in Chapter 4 (Section 4). An analogous variational principle is applicable for the case of convective heat transfer. It is obtained by putting equal to zero the variation

$$\tfrac{1}{2}\delta\iiint_\tau \frac{1}{k}\mathbf{\Theta}'^2\,d\tau + \int_A\int_A r(P', P)\Theta'_n(P')\,\delta\Theta'_n(P)\,dA_P\,dA_{P'} = 0 \tag{4.24}$$

under the condition that

$$\operatorname{div}\mathbf{\Theta}' = -c\theta. \tag{4.25}$$

The variable t does not appear explicitly because the fields involve instantaneous values at a given time. The field θ is given and the variations $\delta\mathbf{\Theta}'$ must satisfy the constraint (4.25). The proof follows exactly the same procedure as in Section 4 of Chapter 4.

As shown below, the function $r(P', P)$ is related to the trailing function

$\bar{r}(P, P')$ of the *adjoint system* obtained by reversing the velocity field in the fluid.

Reverse flow theorem

Consider a homogeneous isotropic incompressible fluid in steady flow. The temperature field satisfies the equation

$$\operatorname{div}(cA \operatorname{grad} \theta) = c\mathbf{u} \operatorname{grad} \theta. \qquad (4.25\,\text{a})$$

The flow of velocity $\mathbf{u}$ may be laminar or turbulent and the value of A is written

$$A = \frac{k}{c} + \epsilon, \qquad (4.25\,\text{b})$$

where ϵ is the turbulent diffusivity assumed to be isotropic. It may be a function of the coordinates, while c and k are constant. The temperature $\bar{\theta}$ in the same fluid when we reverse the velocity field *without changing the diffusivity* ϵ, is governed by the equation

$$\operatorname{div}(cA \operatorname{grad} \bar{\theta}) = -c\mathbf{u} \operatorname{grad} \bar{\theta}). \qquad (4.25\,\text{c})$$

If we take into account the condition of incompressibility,

$$\operatorname{div} \mathbf{u} = 0, \qquad (4.25\,\text{d})$$

equations (4.25 a) and (4.25 c) lead to the relation

$$\operatorname{div}(cA\bar{\theta} \operatorname{grad} \theta - cA\theta \operatorname{grad} \bar{\theta} - c\theta\bar{\theta}\mathbf{u}) = 0. \qquad (4.25\,\text{e})$$

We integrate this expression over a large volume of the fluid whose boundary is limited by a surface S of the fluid and a solid wall W. The volume integral may be transformed into a surface integral over the boundary $S+W$. If we assume that the temperature field θ vanishes at large distance, the surface integral over S vanishes. The integral is then reduced to a surface integral over the solid wall. If we take into account the condition that the normal component of the fluid velocity $\mathbf{u}$ vanishes at the wall, we obtain

$$\iint_W (cA\bar{\theta} \operatorname{grad}_n \theta - cA\theta \operatorname{grad}_n \bar{\theta})\, dW = 0. \qquad (4.25\,\text{f})$$

Assume now that the field θ is produced by a unit rate of concentrated heat injection into the fluid at the point P' of the wall. We may write

$$cA \operatorname{grad}_n \theta = -\delta(P, P'), \qquad (4.25\,\text{g})$$

where $\delta(P, P')$ is a Dirac function. The surface temperature θ at point P of the surface represents the corresponding trailing function

$$\theta = r(P, P'). \qquad (4.25\,\text{h})$$

Similarly, $\bar{\theta}$ may be chosen to represent the trailing function due to a heat injection at point P'' in the reverse flow, Hence

$$cA \operatorname{grad}_n \bar{\theta} = -\delta(P, P''), \qquad \bar{\theta} = \bar{r}(P, P''). \qquad (4.25\,\text{i})$$

By substituting the values (4.25 g), (4.25 h), and (4.25 i) into the surface integral (4.25 f), we derive

$$\iint_W \{\bar{r}(P, P'')\delta(P, P') - r(P, P')\delta(P, P'')\}\, dW_P = 0. \qquad (4.25\,\text{j})$$

Hence

$$\bar{r}(P', P'') = r(P'', P'). \qquad (4.25\,\text{k})$$

This result establishes that the trailing function for reverse flow is obtained by interchanging the points in the trailing function for direct flow.

5. UNIFIED EQUATIONS FOR FLUID–SOLID SYSTEMS WITH CONVECTION

In the previous sections it was shown how the problem of heat conduction could be formulated by Lagrangian equations in a solid with boundaries in contact with a moving fluid. We shall now derive a still more general result, which provides a unified formulation by Lagrangian equations of heat conduction and convection in a composite system that includes solids and moving fluids.

We shall assume the fluids to be incompressible. In practice, this amounts to stating that the influence of compressibility may be neglected. However, the nature of the fluids may otherwise be quite general and includes the case of non-homogeneity.

The initial step generalizing the Lagrangian equations to heat convection in a moving fluid was provided by Nigam and Agrawal‡ and the author.§ The more general treatment given here was developed in a subsequent paper.||

Heat content and thermal potential of a fluid particle

In order to include non-homogeneous fluids, hence fluids composed of particles that are physically different, we must extend some previously defined concepts to individual fluid particles. We identify the particles by their initial coordinates X_i at the time $t = 0$. The coordinates x_i of these particles at the time t are

$$x_i = x_i(X, t), \tag{5.1}$$

where X denotes the three initial coordinates. The fluid velocity field is

$$v_i = \frac{\partial}{\partial t} x_i(X, t). \tag{5.2}$$

Under the assumption of incompressibility it is possible to define a heat capacity c of a fluid particle per unit volume as

$$c = c(X, \theta). \tag{5.3}$$

It is a function of the temperature θ. The dependence of this value on the physical nature of the fluid particle is expressed by considering c to depend not only on the temperature θ but also on the initial coordinates

‡ S. D. Nigam and H. C. Agrawal, 'A variational principle for convection of heat', *J. Math. Mech.* **9**, 869–84 (1960).

§ M. A. Biot, 'Lagrangian thermodynamics of heat transfer in systems including fluid motion', *J. Aerospace Sci.* **29**, 568–77 (1962).

|| M. A. Biot, 'Generalized variational principles for convective heat transfer and irreversible thermodynamics', *J. Math. Mech.* **15**, 177–86 (1966).

X. These coordinates are used here in order to 'tag' the individual particles.

Two other fundamental quantities may be derived from the foregoing definition of c. They are:

the *heat content* of a particle per unit volume,

$$h(X,\theta) = \int_0^\theta c(X,\theta)\, d\theta; \tag{5.4}$$

the *thermal potential* of a particle per unit volume,

$$F(X,\theta) = \int_0^\theta \theta\, dh, \tag{5.5}$$

where $dh = c\,d\theta$.

In the following derivations we shall express the physical quantities in terms of the time and fixed coordinates, x_i. Equations (5.1), when solved for X_i, become

$$X_i = X_i(x,t), \tag{5.6}$$

which determine the initial coordinates of a particle as functions of its coordinates at the time t. By substituting the values (5.6) of the initial coordinates into expressions (5.2), (5.3), (5.4), and (5.5) they become

$$\begin{aligned} v_i &= v_i(x,t), & c &= c(x,t,\theta), \\ h &= h(x,t,\theta), & F &= F(x,t,\theta). \end{aligned} \tag{5.7}$$

Fundamental physical laws

The physics of thermal diffusion and convection is governed by two basic equations,

$$\dot{h} + \sum^{i} \frac{\partial}{\partial x_i}(J_i + v_i h) = 0 \tag{5.8}$$

and

$$\frac{\partial \theta}{\partial x_i} = -\sum^{j} \lambda_{ij} J_j \tag{5.9}$$

(incompressibility implies $\sum^{i} \partial v_i/\partial x_i = 0$). Equation (5.8), where $\dot{h} = \partial h/\partial t$, expresses conservation of energy. The vector J_i represents the rate of flow of thermal energy *relative to the fluid material*, while $v_i h$ is the convective rate of flow. Equations (5.9) express the law of heat conduction. They coincide with equations (5.6) of Chapter 1, where $\lambda_{ij} = \lambda_{ji}$ is the thermal resistivity tensor. It may depend on x, t, and θ.

An essential feature here results from the use of a heat-displacement vector field, $H_i(x,t)$, to define h and J_i by the expressions

$$h = -\sum^{i} \frac{\partial H_i}{\partial x_i}, \qquad J_i = \dot{H}_i - v_i h. \tag{5.10}$$

The purpose of this representation is to satisfy *identically* the conservation equation (5.8). Hence, by using H_i as the unknown physical field, we need only to satisfy the thermal conduction equation (5.9).

An important aspect of the present analysis is the fact that the conservation equation (5.8) behaves as an *holonomic constraint* in analogy with classical mechanics. Attention is called to the physical significance of the vector

$$\dot{H}_i = J_i + v_i h. \tag{5.11}$$

It represents the total conductive and convective rate of heat flow per unit area through a surface *fixed in space.*

Variational principle and Lagrangian equations

We write the variational equation

$$\sum^{i}\left(\frac{\partial\theta}{\partial x_i}+\sum^{j}\lambda_{ij}J_j\right)\delta H_i = 0. \tag{5.12}$$

For arbitrary variations δH_i, this equation is equivalent to equations (5.9) for heat conduction. Equation (5.12) is integrated over a volume τ and the first term is integrated by parts, using the relations

$$\delta h = -\sum^{i}\frac{\partial}{\partial x_i}\delta H_i, \qquad \delta F = \theta\,\delta h, \tag{5.13}$$

which are consequences of equations (5.5) and (5.10). We obtain

$$\iiint_\tau \delta F\,d\tau + \iiint_\tau \sum^{ij}\lambda_{ij}J_j\,\delta H_i\,d\tau = -\iint_A \theta\sum^{i} n_i\,\delta H_i\,dA. \tag{5.14}$$

The surface integral is extended to the boundary A of the volume τ, and n_i denotes the unit outward normal to this boundary. We introduce the total thermal potential V in the volume τ as

$$V = \iiint_\tau F\,d\tau. \tag{5.15}$$

With this definition equation (5.14) is written

$$\delta V + \iiint_\tau \sum^{ij}\lambda_{ij}J_j\,\delta H_i\,d\tau = -\iint_A \theta\sum^{i} n_i\,\delta H_i\,dA. \tag{5.16}$$

This results generalizes the variational principle to systems that include convection. Its form is quite similar to the variational principle for pure conduction as given by equations (5.10) of Chapter 1.

As before, the principle may be translated into Lagrangian equations by considering the vector field H_i to be a function of n generalized coordinates q_i. We write

$$H_i = H_i(q_1, q_2, \ldots, q_n, x_1, x_2, x_3, t). \tag{5.17}$$

The variations δH_i are due entirely to the variations of q_i. They become

$$\delta H_i = \sum^{j} \frac{\partial H_i}{\partial q_j} \delta q_j. \tag{5.18}$$

Furthermore,
$$\dot{H}_i = \sum^{j} \frac{\partial H_i}{\partial q_j} \dot{q}_j + \frac{\partial H_i}{\partial t}. \tag{5.19}$$

This leads to the relation
$$\frac{\partial \dot{H}_i}{\partial \dot{q}_j} = \frac{\partial H_i}{\partial q_j}. \tag{5.20}$$

Hence the variations (5.18) may be written as

$$\delta H_i = \sum^{j} \frac{\partial \dot{H}_i}{\partial \dot{q}_j} \delta q_j. \tag{5.21}$$

By introducing the values (5.18) and (5.21) for δH_i into the variational principle (5.16), we derive the Lagrangian equations

$$\frac{\partial V}{\partial q_i} + \frac{\partial D}{\partial \dot{q}_i} = Q_i. \tag{5.22}$$

The dissipation function in this case is defined as

$$D = \tfrac{1}{2} \iiint_{\tau} \sum^{ij} \lambda_{ij} J_i J_j \, d\tau. \tag{5.23}$$

We must assume here again that the thermal resistivity is symmetric ($\lambda_{ij} = \lambda_{ji}$) in accordance with the Onsager principle. The thermal force Q_i is defined as before as

$$Q_i = -\iint_{A} \theta \sum^{j} n_j \frac{\partial H_j}{\partial q_i} \, dA. \tag{5.24}$$

Turbulent flow

The results presented here are applicable to a homogeneous incompressible fluid in turbulent flow, by introducing the tensor

$$K_{ij} = k\,\delta_{ij} + c\epsilon_{ij}. \tag{5.25}$$

In this expression δ_{ij} is the Kronecker symbol, k and c represent respectively the thermal conductivity and heat capacity of the fluid, and ϵ_{ij} is the turbulent diffusivity tensor. The turbulent diffusivity also satisfies the reciprocity relations

$$\epsilon_{ij} = \epsilon_{ji}. \tag{5.26}$$

We now define a total resistivity matrix

$$[\lambda_{ij}] = [K_{ij}]^{-1}, \tag{5.27}$$

where the right side denotes the inverse of the matrix $[K_{ij}]$. With this definition of λ_{ij} the dissipation function (5.23) and the Lagrangian equations (5.22) remain valid for turbulent flow.

Internal heat sources

In many problems it is necessary to take into account internal generation of heat in the fluid. For example, the heat may be generated by viscous friction, chemical reactions, nuclear reactions, or radiation absorption. The effect of thermal source may be included by generalizing the procedure outlined in Chapter 1, Section 6. We denote by

$$w = w(x_1, x_2, x_3, t) \tag{5.28}$$

the heat generated per unit volume and unit time. The energy conservation equation (5.8) is now written as

$$\dot{h} + \sum^{i} \frac{\partial}{\partial x_i}(J_i + v_i h) = w. \tag{5.29}$$

This equation is satisfied identically by putting

$$h = -\sum^{i} \frac{\partial H_i}{\partial x_i}, \qquad J_i = \dot{H}_i - v_i h + \dot{H}_i^*, \tag{5.30}$$

where H_i^* is chosen so as to satisfy the relation

$$\sum^{i} \frac{\partial H_i^*}{\partial x_i} = \int_0^t w\, dt. \tag{5.31}$$

With these definitions, Lagrangian equations are derived by following the same procedure as for the case of pure conduction described in Chapter 1, Section 6.

Application to mixed solid–fluid systems

The Lagrangian equations derived in this section provide a unified formulation of conduction and convection in mixed systems composed of solids and moving fluids. The domain of integration may contain both solids and fluids. In the solid portion the velocity v_i is put equal to zero and λ_{ij} represents the thermal resistivity of the solid. In the fluid portion λ_{ij}, as defined above, includes the effect of turbulent diffusivity.

This unified formulation should be particularly applicable to heat transfer in a porous solid containing a moving fluid.

Principle of minimum dissipation for convective heat transfer

Equations (5.22) are identical in form to the Lagrangian equations (4.6) of Chapter 1. The principle of minimum dissipation derived in that chapter depends only on the form of these equations and is therefore applicable to convective heat transfer provided the dissipation function is defined by expression (5.23).

CHAPTER SEVEN

BOUNDARY-LAYER HEAT TRANSFER

1. INTRODUCTION

THE present chapter deals with the variational evaluation of the trailing function and its application to laminar and turbulent boundary layers. The material presented here was developed initially in two papers by the author.‡§ The trailing function provides the key to a simple evaluation of convective heat transfer. At the same time it also embodies a simple physical model for the understanding of the basic mechanism of the combined conduction and convection in a moving fluid.

As shown in Section 2, the variational methods and techniques developed for thermal conduction are readily applicable to boundary-layer heat transfer by the use of a conduction analogy. This provides a convenient procedure for the evaluation of the trailing function by variational methods, as illustrated in a simple example in Section 3. The example also shows that the method is remarkably accurate even when using a crude approximation for the temperature distribution in the fluid.

In Section 4 a general procedure is described for the variational evaluation of the trailing function for boundary layers. Fundamental equations and parameters are derived in non-dimensional form. The analysis is carried out numerically in Section 5 for laminar boundary layers. An approximate piece-wise analytical expression is derived for the trailing function in non-dimensional form. It is found that this expression is not sensitive to variations in the velocity profile and is therefore applicable to a large number of cases. A comparison with an exact result again confirms the accuracy of the variational procedure. A similar piece-wise analytical expression for the trailing function is obtained in Section 6 for the turbulent boundary layer. Due to an appropriate choice of parameters, the numerical results obtained here may be considered as typical for boundary-layer heat transfer. In

‡ M. A. Biot, 'Lagrangian thermodynamics of heat transfer in systems including fluid motion', *J. Aerospace Sci.* **29**, 568–77 (1962).

§ M. A. Biot, 'Simplified variational and physical analysis of heat transfer in laminar and turbulent flow'. *Physics Fluids*, **10**, 1424–37 (1967).

particular, it is important to introduce a special definition of the characteristic boundary-layer thickness.

How the concept of trailing function may be used to formulate general problems of boundary-layer heat transfer is illustrated in Section 7. The problem of heat transfer for the free-stream boundary layer is considered. It is also shown how the concept of trailing function must be slightly modified in the case of ducted flow. Finally, the problem of heat transfer is formulated for heat exchangers with two fluids flowing in opposite directions.

The practical importance of the trailing function results from its fundamental physical significance. It embodies the main heat-transfer properties of the fluid as a quasi-local phenomenon in a restricted downstream region from a given point. This provides an intuitive approach well suited to variational analysis.

Special emphasis has also been put on the Peclet number as a fundamental physical parameter that measures the interaction between conduction and convection in the fluid.

2. CONDUCTION ANALOGY

In many problems of convective heat transfer in a fluid it is possible to transform the equations to a form that coincides with the case of pure conduction. This has the advantage of simplifying considerably the application of variational principles.

Consider a stationary two-dimensional flow of an incompressible fluid. For constant heat capacity and thermal conductivity the temperature field θ satisfies the following equation:

$$cu\frac{\partial\theta}{\partial x}+cv\frac{\partial\theta}{\partial y} = k\left(\frac{\partial^2\theta}{\partial x^2}+\frac{\partial^2\theta}{\partial y^2}\right). \tag{2.1}$$

The velocity field components are

$$u = u(x,y), \qquad v = v(x,y), \tag{2.2}$$

and the property of incompressibility is expressed by the condition

$$\frac{\partial u}{\partial x}+\frac{\partial v}{\partial y} = 0. \tag{2.3}$$

A standard procedure in convective heat-transfer analysis is to neglect the term $\partial^2\theta/\partial x^2$, which represents thermal conduction in the direction of flow.‡ This leads to the simplified equation,

$$cu\frac{\partial\theta}{\partial x}+cv\frac{\partial\theta}{\partial y} = k\frac{\partial^2\theta}{\partial y^2}. \tag{2.4}$$

A similar equation is obtained for the case of *turbulent flow*, by introducing an 'effective thermal conductivity',

$$k' = k+c\epsilon(x,y). \tag{2.5}$$

In this expression $\epsilon(x,y)$ is the eddy diffusivity due to turbulence. The diffusivity is assumed to be isotropic. Since ϵ is a function of the coordinates, k' is not constant and equation (2.4) is replaced by

$$cu\frac{\partial\theta}{\partial x}+cv\frac{\partial\theta}{\partial y} = \frac{\partial}{\partial y}\left(k'\frac{\partial\theta}{\partial y}\right). \tag{2.6}$$

Laminar flow is included as a particular case when $\epsilon = 0$. Further simplification of equation (2.6) was recently derived by the author§ by the following procedure, which generalizes the von Mises transformation.||

We introduce a change of variables:

$$x' = x'(x,y), \qquad y' = y'(x,y). \tag{2.7}$$

The variables x' and y' may be considered as curvilinear coordinates. Let us choose the function y' in such a way that its value is constant along the streamlines. By definition,

$$u\frac{\partial y'}{\partial x}+v\frac{\partial y'}{\partial y} = 0. \tag{2.8}$$

Also
$$\frac{\partial\theta}{\partial x} = \frac{\partial\theta}{\partial x'}\frac{\partial x'}{\partial x}+\frac{\partial\theta}{\partial y'}\frac{\partial y'}{\partial x}, \qquad \frac{\partial\theta}{\partial y} = \frac{\partial\theta}{\partial x'}\frac{\partial x'}{\partial y}+\frac{\partial\theta}{\partial y'}\frac{\partial y'}{\partial y}. \tag{2.9}$$

We now choose a particular case of the transformation (2.9) by putting

$$x' = x, \qquad \frac{\partial x'}{\partial x} = 1, \qquad \frac{\partial x'}{\partial y} = 0. \tag{2.10}$$

By combining equations (2.8), (2.9), and (2.10) we obtain

$$u\frac{\partial\theta}{\partial x}+v\frac{\partial\theta}{\partial y} = u\frac{\partial\theta}{\partial x'}, \qquad \frac{\partial\theta}{\partial y} = \frac{\partial\theta}{\partial y'}\frac{\partial y'}{\partial y}. \tag{2.11}$$

We put
$$\frac{\partial y'}{\partial y} = \frac{1}{\alpha(x',y')}. \tag{2.12}$$

‡ A quantitative evaluation of the range of validity of this approximation was given in the author's paper 'Fundamentals of boundary-layer heat transfer with streamwise temperature variations,' *J. Aerospace Sci.* **25**, 558–67 (1962).

§ See p. 117, n. §.

|| In the von Mises transformation the stream function ψ is used instead of the more general variable y'.

With these results it is possible to write equation (2.6) in the form

$$cu\alpha\frac{\partial\theta}{\partial x'} = \frac{\partial}{\partial y'}\left(\frac{k'}{\alpha}\frac{\partial\theta}{\partial y'}\right). \tag{2.13}$$

We may also write x instead of x', since this is only a change of notation:

$$cu\alpha\frac{\partial\theta}{\partial x} = \frac{\partial}{\partial y'}\left(\frac{k'}{\alpha}\frac{\partial\theta}{\partial y'}\right). \tag{2.14}$$

In this equation α, u, and k'/α are functions of the independent variables x and y'. However, because of the assumption of incompressibility, the product

$$\alpha u = u_r(y') \tag{2.15}$$

depends only on y'. This can be shown by considering two streamlines of coordinates y_1' and y_2'. The integral

$$\int_{y_1}^{y_2} u\,dy = \int_{y_1'}^{y_2'} u\frac{\partial y}{\partial y'}\,dy' = \int_{y_1'}^{y_2'} u\alpha\,dy' \tag{2.16}$$

represents the volume flow between the two streamlines. Since the fluid is incompressible the integral (2.16) is constant. Therefore $u\alpha$ is independent of x along any streamline y', and equation (2.14) may finally be written as

$$cu_r(y')\frac{\partial\theta}{\partial x} = \frac{\partial}{\partial y'}\left(\frac{k'}{\alpha}\frac{\partial\theta}{\partial y'}\right). \tag{2.17}$$

This equation represents a transient conduction problem where x is a time variable and y' is the space variable. Hence the convective heat transfer may be solved by considering an analogue model of one-dimensional heat conduction in a non-homogeneous medium where the heat capacity per unit volume is $cu_r(y')$ and depends on the location, while the thermal conductivity k'/α depends on both time and location. This establishes a *conduction* analogy.

As a consequence, the variational methods derived for heat conduction are applicable to convective heat transfer in laminar and turbulent flow.

3. VARIATIONAL EVALUATION OF THE TRAILING FUNCTION

Variational methods combined with the conduction analogy provide a remarkably accurate and simple method for the evaluation of the trailing function in convective heat-transfer.

Our purpose in this section is to illustrate the procedure by considering a very simple case. We assume a uniform laminar flow with constant velocity U. The x-axis is located in the plane boundary and the velocity is oriented along x. The y-axis is normal to the boundary. We are dealing here with a two-dimensional flow field in the x, y plane of velocity components:

$$u = U, \qquad v = 0. \tag{3.1}$$

The initial temperature of the fluid is assumed to be $\theta = 0$.

At the boundary we consider a straight line through the origin and normal to the x-axis. At this line heat is injected into the fluid at a rate equal to unity per unit length. No heat flow occurs elsewhere across the boundary. This generates a down-stream temperature distribution along x equal to

$$\theta = r(x). \tag{3.2}$$

If the injection line is located at the abscissa ξ instead of the origin, the down-stream temperature would be

$$\theta = r(x-\xi). \tag{3.3}$$

By putting $x = s$ and $\xi = s'$, we may write

$$\theta = r(s-s') = r(s,s'). \tag{3.4}$$

Hence $r(s-s')$ represents a particular case of the one-dimensional trailing function defined by expression (2.9) of Chapter 6. We have assumed here that the adiabatic temperature of the fluid is $\theta_a = 0$. This does not restrict the generality of the results if the principle of superposition is applicable. Since the flow is laminar we put

$$k' = k. \tag{3.5}$$

With the values (3.1) for the velocity field, equation (2.4) becomes

$$cU\frac{\partial \theta}{\partial x} = k\frac{\partial^2 \theta}{\partial y^2}. \tag{3.6}$$

We assume that the injection line is located at the origin ($\xi = 0$). The problem is to evaluate the trailing function $r(x)$. Since we are dealing with a two-dimensional problem we consider a slab of fluid of unit thickness, parallel to the xy plane. The rate of heat injection in this slab at the origin is then equal to unity.

We integrate both sides of equation (3.6) with respect to y from $y = 0$ to $y = \infty$ and we obtain

$$\frac{\partial I}{\partial x} = -k\left(\frac{\partial \theta}{\partial y}\right)_{y=0}, \tag{3.7}$$

where

$$I(x) = \int_0^\infty cU\theta\, dy. \tag{3.8}$$

In deriving this result it is assumed that $\partial\theta/\partial y$ vanishes for $y = \infty$.

Expression $-k(\partial\theta/\partial y)$ represents the rate of heat injection into the fluid. Hence we may write

$$\frac{\partial I}{\partial x} = -k\left(\frac{\partial \theta}{\partial y}\right)_{y=0} = \delta(x), \tag{3.9}$$

where $\delta(x)$ vanishes except in the arbitrarily small interval $-\epsilon < x < \epsilon$, while the following integral remains verified:

$$\int_{-\epsilon}^{+\epsilon} \delta(x)\, dx = 1. \tag{3.10}$$

Such a 'Dirac function' $\delta(x)$ expresses the property that a unit rate of heat injection is concentrated at $x = 0$.

We now consider the function $I(x)$. By integrating equation (3.9) with respect to x we derive

$$I(x) = \int_{-\epsilon}^{x} \delta(x)\, dx = 1. \tag{3.11}$$

Note that $I(-\epsilon) = 0$ because θ vanishes up-stream from the point of injection. Hence, according to relations (3.8) and (3.11),

$$\int_{0}^{\infty} cU\theta\, dy = 1 \quad \text{for } x > 1. \tag{3.12}$$

These results may be interpreted in terms of a conduction analogy. Equation (3.6) represents the one-dimensional thermal conduction in a solid of heat capacity cU and thermal conductivity k with the time variable $t = x$. The solid occupies the region $y > 0$ with a plane boundary at $y = 0$. A unit amount of heat is injected per unit area at the boundary $y = 0$ at the time $t = 0$ and the boundary is immediately sealed. The heat is then allowed to diffuse into the solid. The total amount of heat remains constant and equal to unity according to equation (3.12).

In order to apply the variational method to this analogue problem we assume a temperature distribution

$$\theta = \frac{3}{2cUq}\left(1-\frac{y^2}{q^2}\right) \quad (y < q). \tag{3.13}$$

It satisfies the constraint represented by the constant value of the integral (3.12). The thermal potential of the analogue model is

$$V = \tfrac{1}{2}cU\int_{0}^{q} \theta^2\, dy = \frac{3}{5cUq}. \tag{3.14}$$

The value of H is

$$H = cU\int_{y}^{q} \theta\, dy = 1-\frac{3}{2}\frac{y}{q}+\frac{1}{2}\frac{y^3}{q^3} \tag{3.15}$$

and the dissipation function is

$$D = \frac{1}{2k}\int_0^q \dot{H}^2\,dy = \frac{3}{35k}\frac{\dot{q}^2}{q}. \tag{3.16}$$

In this expression $\dot{q} = dq/dx$ since x plays the role of the time variable. The thermal force Q is equal to zero since the boundary is insulated. Mathematically this is expressed by the property $\delta H = 0$ at $y = 0$. Hence the Lagrangian equation is

$$\frac{\partial V}{\partial q} + \frac{\partial D}{\partial \dot{q}} = 0. \tag{3.17}$$

Substitution of the values (3.14) and (3.16) yields the differential equation

$$2q\dot{q} = 7\frac{k}{cU}. \tag{3.18}$$

By integration with the initial condition, $q = 0$ for $x = 0$, we derive

$$q = \sqrt{\left(\frac{7kx}{cU}\right)}. \tag{3.19}$$

The temperature at the boundary is obtained by putting $y = 0$ in expression (3.13). Its value becomes

$$\theta = \frac{3}{2cUq}. \tag{3.20}$$

Substitution of the value (3.19) for q yields

$$\theta = \frac{1}{\sqrt{\{(28/9)cUkx\}}}. \tag{3.21}$$

This temperature represents the trailing function for heat injection at $\xi = 0$. We write it as

$$r(x) = \frac{1}{\sqrt{\{(28/9)cUkx\}}}. \tag{3.22}$$

Since the flow field is independent of x the trailing function for heat injection at the abscissa ξ is $r(x-\xi)$.

We may compare this result with the exact solution. Equation (3.6) is verified by the solution

$$\theta = \frac{1}{\sqrt{(\pi cUkx)}}\exp\left(-\frac{cUy^2}{4kx}\right). \tag{3.23}$$

This expression also satisfies the condition of constant value of the integral (3.12). The exact value of the trailing function obtained by putting $y = 0$ in expression (3.23) is therefore

$$r(x) = \frac{1}{\sqrt{(\pi c U k x)}}. \tag{3.24}$$

Comparing with the approximate value (3.22) we can see that the constant, $\sqrt{(28/9)} = 1{\cdot}765$, is replaced by $\sqrt{\pi} = 1{\cdot}772$. Hence the error of the approximate value is less than $\frac{1}{2}$ per cent.

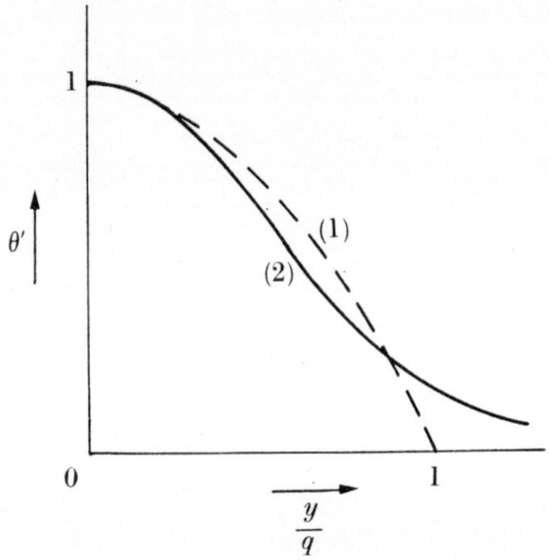

FIG. 7.1. Value of $\theta' = \theta\sqrt{(\pi c U k x)}$ as a function of y/q. (1) Parabolic approximation (3.13). (2) Exact gaussian distribution (3.23).

We may also compare the approximate and exact temperature distribution (3.13) and (3.23) inside the fluid. They are plotted in non-dimensional form as $\theta' = \theta\sqrt{(\pi c U k x)}$ in Fig. 7.1. It is interesting to note that high accuracy is obtained for the trailing function even with a somewhat crude approximation for the temperature distribution in the fluid.

4. GENERAL VARIATIONAL PROCEDURES

Knowledge of the trailing function provides the key to the evaluation of boundary-layer heat transfer. The application of the trailing function to typical problems of boundary-layer heat transfer will be discussed below in Section 7.

General procedures for the variational evaluation of the trailing function will now be derived by introducing certain standardized approximations along with non-dimensional parameters and variables that reflect fundamental physical properties.

We shall first assume parallel streamlines and show how the method may be extended to the case of non-parallel streamlines. In this analysis we shall put $\theta_a = 0$. Because the principle of superposition is assumed to be valid, this does not restrict the generality of the results.

Parallel streamlines and laminar flow

The conduction analogy for this case is obtained by putting $v = 0$ in equation (2.4). We derive

$$cu\frac{\partial\theta}{\partial x} = k\frac{\partial^2\theta}{\partial y^2}. \tag{4.1}$$

The velocity profile of the boundary layer is

$$u = u(y). \tag{4.2}$$

By putting

$$c'(y) = cu(y) \tag{4.3}$$

equation (4.1) becomes

$$c'(y)\frac{\partial\theta}{\partial x} = k\frac{\partial^2\theta}{\partial y^2}. \tag{4.4}$$

The trailing function is obtained by assuming a unit rate of heat injection at $x = 0$. We follow the same procedure as in Section 3. In the present case the integral condition (3.12) is replaced by

$$\int\limits_0^\infty c'(y)\theta\, dy = 1 \quad \text{for } x > 0. \tag{4.5}$$

Equations (4.4) and (4.5) represent a conduction analogy in a medium with non-uniform heat capacity $c'(y)$ and a condition of total heat content equal to unity. The conduction analogy may be formulated in non-dimensional form by putting

$$u(y) = U\varphi(\eta) \tag{4.6}$$

with

$$\eta = \frac{y}{\delta}. \tag{4.7}$$

We denote by U a reference velocity and by δ a characteristic thickness of the boundary layer. We also introduce a non-dimensional variable,

$$\tau = \frac{1}{Pe}\frac{x}{\delta}, \tag{4.8}$$

where

$$Pe = \frac{cU\delta}{k} \tag{4.9}$$

is the Peclet number derived from U and δ.

With these definitions equation (4.4) is written in non-dimensional form:

$$\varphi(\eta)\frac{\partial\theta}{\partial\tau} = \frac{\partial^2\theta}{\partial\eta^2}. \tag{4.10}$$

The conduction analogy governed by this equation is represented by transient one-dimensional heat conduction in a medium of heat capacity $c = \varphi(\eta)$ function of the depth coordinate η with thermal conductivity $k = 1$. The time is represented by the variable τ. The integral condition (4.5) becomes

$$\int_0^\infty \theta\varphi(\eta)\,d\eta = H_0 \tag{4.11}$$

with

$$H_0 = \frac{1}{kPe}. \tag{4.12}$$

Hence in the non-dimensional conduction analogy, H_0 represents the amount of heat injected at the time $\tau = 0$.

The variational principle is now applied to derive the trailing function corresponding to heat injection at $x = 0$, using the conduction analogy corresponding to equation (4.10). The temperature distribution in the semi-infinite solid in the region $\eta > 0$ is approximated by the cubic,

$$\begin{aligned} \theta &= \theta_0\left(1-\frac{\eta^3}{q^3}\right) && \text{for } \eta < q, \\ \theta &= 0 && \text{for } \eta > q, \end{aligned} \tag{4.13}$$

where q is a penetration depth considered to be an unknown function of the time τ. A cubic is chosen instead of the parabolic approximation (3.13) because it yields more accurate solutions for the usual velocity profiles where $\varphi(\eta)$ is proportional to η in the vicinity of the boundary $\eta = 0$. The temperature θ_0 at $\eta = 0$ in expression (4.13) is a function of q determined by the integral relation (4.11). This functional relation is obtained by substituting the value (4.13) of θ into the integral (4.11). It is written as

$$\theta_0\left(A-\frac{B}{q^3}\right) = H_0, \tag{4.14}$$

with the following functions of q:

$$\begin{aligned} A &= \int_0^q \varphi(\eta)\,d\eta, \\ B &= \int_0^q \eta^3\varphi(\eta)\,d\eta. \end{aligned} \tag{4.15}$$

The thermal potential corresponding to equation (4.10) is

$$V = \tfrac{1}{2}\int_0^q \varphi(\eta)\theta^2\,d\eta. \tag{4.16}$$

The analogue heat displacement at the point η is

$$H = \int_{\eta}^{q} \theta\varphi(\eta)\, d\eta. \tag{4.17}$$

It is convenient to transform this expression by using the integral condition (4.11) and write it in the form

$$H = H_0 - \int_{0}^{\eta} \theta\varphi(\eta)\, d\eta. \tag{4.18}$$

We derive

$$\dot{H} = -\int_{0}^{\eta} \dot{\theta}\varphi(\eta)\, d\eta. \tag{4.19}$$

Evaluation of this expression requires the use of the relation

$$\frac{\dot{\theta}_0}{\theta_0} = -R\frac{\dot{q}}{q}, \tag{4.20}$$

where

$$R = \frac{3B}{Aq^3 - B} \tag{4.21}$$

is a function of q. Relation (4.20) is obtained by differentiating equation (4.14) with respect to τ. The dissipation function D is therefore

$$D = \tfrac{1}{2}\int_{0}^{q} \dot{H}^2\, d\eta = \tfrac{1}{2}\theta_0^2\, Mq\dot{q}^2 \tag{4.22}$$

with M a function of q.

The Lagrangian equation for q is

$$\frac{\partial V}{\partial q} + \frac{\partial D}{\partial \dot{q}} = 0. \tag{4.23}$$

The right side is zero because no flow is allowed across the boundary at $\eta = 0$ and the corresponding thermal force Q vanishes. From equation (4.16) we also derive

$$\frac{\partial V}{\partial q} = -\theta_0^2\, L, \tag{4.24}$$

with L a function of q.

By introducing expressions (4.22) and (4.24) into the Lagrangian equation (4.23) we obtain

$$g(q)q\dot{q} = 1, \tag{4.25}$$

where

$$g(q) = \frac{M}{L}. \tag{4.26}$$

With the initial condition $q = 0$ for $\tau = 0$, the integral of the differential equation (4.25) is

$$\tau = \int_0^q qg(q)\,dq. \tag{4.27}$$

On the other hand, equation (4.14) yields θ_0 as a function of q:

$$\theta_0 = \frac{H_0}{A-B/q^3}. \tag{4.28}$$

From equations (4.27) and (4.28) it is possible to plot θ_0 as a function of τ by evaluating the ordinates and abscissae as parametric functions of q. The resulting plot is represented by the relation

$$\phi(\tau) = \frac{\theta_0}{H_0} = \frac{1}{A-B/q^3}, \tag{4.29}$$

where $\phi(\tau)$ is a non-dimensional form of the trailing function, introduced by the author under the name *reduced trailing function.*

The value θ_0 represents the actual temperature at the boundary. Hence, going back to the physical variables through equations (4.8) and (4.12), the trailing function is

$$\theta_0 = r(x) = \frac{1}{kPe}\phi\left(\frac{1}{Pe}\frac{x}{\delta}\right). \tag{4.30}$$

Parallel streamlines and turbulent flow

It will now be shown that the non-dimensional form of the conduction analogy represented by equation (4.10) may be extended to turbulent boundary layers with parallel streamlines. This case is obtained by putting

$$\begin{aligned} y' &= y, \qquad \alpha = 1, \\ u_r &= u(y), \\ k'(y) &= k+c\epsilon(y), \end{aligned} \tag{4.31}$$

where $\epsilon(y)$ is the eddy diffusivity. With these values equation (2.17) becomes

$$cu(y)\frac{\partial\theta}{\partial x} = \frac{\partial}{\partial y}\left\{k'(y)\frac{\partial\theta}{\partial y}\right\}. \tag{4.32}$$

We denote by U a reference velocity and by δ a characteristic thickness of the boundary layer. We now introduce the same non-dimensional variables, τ and η, as defined above by relations (4.7) and (4.8). With these variables the differential equation (4.32) becomes

$$\varphi(\eta)\frac{\partial\theta}{\partial\tau} = \frac{\partial}{\partial\eta}\left\{\sigma(\eta)\frac{\partial\theta}{\partial\eta}\right\}, \tag{4.33}$$

where
$$\sigma(\eta) = \frac{k'}{k} = 1+\frac{c\epsilon}{k}. \tag{4.34}$$

By the change of variable

$$\eta' = \int_0^\eta \frac{d\eta}{\sigma}, \tag{4.35}$$

equation (4.33) is further simplified to

$$\beta(\eta')\frac{\partial\theta}{\partial\tau} = \frac{\partial^2\theta}{\partial\eta'^2} \tag{4.36}$$

with
$$\beta(\eta') = \varphi(\eta)\sigma(\eta). \tag{4.37}$$

Equation (4.36) is of the same form as equation (4.10) and again represents a non-dimensional conduction analogy with a unit thermal conductivity and a heat capacity $\beta(\eta')$. However, the behaviour of the function $\beta(\eta')$ is quite different from that of $\varphi(\eta)$. The variational procedure outlined for the laminar case and based on the approximate temperature distribution (4.13) is only partially applicable in the case of turbulent flow. A suitably modified procedure will be developed in Section 6 dealing with the numerical analysis of the trailing function in turbulent boundary layers.

Non-parallel streamlines

In this case the trailing function depends on the point of heat injection. We choose this point as origin, hence the abscissa x represents the distance down-stream from this point. Consider the conduction analogy represented by equation (2.17). The coordinate y', which is constant along the streamlines, may be defined in such a way that $y = y'$, $\alpha = 1$ on the y-axis. Hence y is simply the distance along the normal through the point of injection and the function $u_r(y') = \alpha u$ is the velocity profile at this point.

The trailing function for non-parallel streamlines may be obtained by applying the variational analysis to equation (2.17). In general, of course, the factor k'/α in this equation will depend on both y' and x, leading to an analysis that is different from the foregoing procedure for parallel streamlines. However, the latter procedure may still be used if we assume that the main heat-transfer process is restricted to a sufficiently narrow range down-stream from the point of injection. In that case, we may assume that in this range k' and α are nearly independent of x, so that we may put $\alpha = 1$ and

$$k'(x, y') = k'(y'). \tag{4.38}$$

Hence equation (2.17) becomes identical with equation (4.32) for parallel streamlines. These assumptions provide a simple result that may be looked upon as a first approximation for the case of non-parallel streamlines. Note that in the present case U will depend on the point of injection so that Pe represents a *local* Peclet number.

Further refinement of such a first approximation may be derived by applying the variational procedure to equation (2.17).

5. LAMINAR BOUNDARY LAYER

We shall apply the general procedures outlined in the previous section to an approximate numerical evaluation of the trailing function for laminar boundary layers. We shall consider two extreme cases of velocity profiles. In the first case we shall assume a piece-wise linear profile and in the second a parabolic profile. We shall also assume parallel streamlines.

Piece-wise linear velocity profile

The velocity distribution in this case is

$$\begin{aligned} u &= U\frac{y}{\delta} \quad \text{for } y < \delta, \\ u &= U \quad\;\; \text{for } y > \delta, \end{aligned} \tag{5.1}$$

as illustrated in Fig. 7.2 (a). The characteristic thickness δ is the ordinate corresponding to the point at which the velocity acquires the constant value U.

For this case the function $\varphi(\eta)$ defined by equation (4.6) becomes

$$\begin{aligned} \varphi(\eta) &= \eta \quad \text{for } \eta < 1, \\ \varphi(\eta) &= 1 \quad \text{for } \eta > 1. \end{aligned} \tag{5.2}$$

The non-dimensional equation (4.10) for the conduction analogy is now written as

$$\eta\frac{\partial\theta}{\partial\tau} = \frac{\partial^2\theta}{\partial\eta^2} \quad (\eta < 1), \qquad \frac{\partial\theta}{\partial\tau} = \frac{\partial^2\theta}{\partial\eta^2} \quad (\eta > 1). \tag{5.3}$$

Following the general procedure, we use the approximate cubic temperature distribution (4.13). We must therefore distinguish two phases, depending on whether the penetration depth q is smaller or larger than unity.

In the first phase where $q < 1$ the heat has not yet penetrated beyond the thickness δ of the boundary layer. In this phase we put $\varphi(\eta) = \eta$. This case is very simple to evaluate. Expression (4.26) becomes

$$g(q) = \tfrac{2}{11}q \tag{5.4}$$

and equations (4.27) and (4.28) yield

$$\tau = \tfrac{2}{33}q^3, \qquad \theta_0 = \frac{10}{3}\frac{H_0}{q^2}. \tag{5.5}$$

Elimination of q between these two equations leads to

$$\theta_0 = 0{\cdot}514 H_0\,\tau^{-\frac{2}{3}}. \tag{5.6}$$

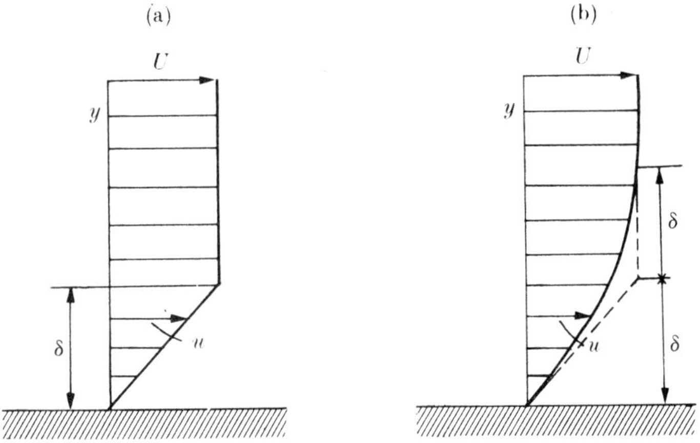

FIG. 7.2. Idealized velocity profiles for laminar boundary layers. (*a*) Piece-wise linear profile. (*b*) Parabolic profile.

Comparison with the general equation (4.29) shows that the reduced trailing function in the first phase is

$$\phi(\tau) = 0{\cdot}514\tau^{-\frac{2}{3}}. \tag{5.7}$$

In the second phase, where $q > 1$, we must evaluate the functions $A(q)$, $B(q)$, and $g(q)$ expressed by equations (4.15) and (4.26) of the general variational procedure. This evaluation must be carried out for the complete range of η from $\eta = 0$ to $\eta = q > 1$, using the two piece-wise linear expressions (5.2). This may be accomplished either numerically or analytically. Furthermore, a plot of the result shows that we may represent the function $g(q)$ by the following analytical approximation:‡

$$\begin{aligned} g(q) &= \tfrac{2}{11}q \qquad \text{for } q < 1, \\ g(q) &= \frac{28q}{73+81q} \quad \text{for } q > 1. \end{aligned} \tag{5.8}$$

This provides a simple evaluation of the integral (4.27) for τ. The value of τ and the value (4.29) of $\phi(\tau)$ are then obtained as functions of q. They are given in Table 7.1 for the second phase ($q > 1$).

‡ See p. 117, n. §.

TABLE 7.1

Reduced trailing function $\phi(\tau)$ for the piece-wise linear velocity profile of Fig. 7.2 (a) in the second phase ($q > 1$) and comparison with the approximation defined by equations (5.9).

q	τ	$\phi(\tau)$	$0{\cdot}514\tau^{-\frac{2}{3}}$	$0{\cdot}554\tau^{-\frac{1}{2}}$
1·00	0·0606	3·33	3·33	
1·50	0·186	1·57	1·58	
2·00	0·386	0·993	0·970	
2·45	0·640	0·750	0·693	0·693
3·00	1·02	0·571		0·548
4·00	1·98	0·400		0·400
6·00	4·91	0·250		0·250

The table also shows that the following piece-wise analytical approximation may be considered satisfactory for most purposes:

$$\begin{aligned} \phi(\tau) &= 0{\cdot}514\tau^{-\frac{2}{3}} \quad \text{for } \tau < 0{\cdot}64, \\ \phi(\tau) &= 0{\cdot}554\tau^{-\frac{1}{2}} \quad \text{for } \tau > 0{\cdot}64. \end{aligned} \tag{5.9}$$

The value $0{\cdot}514\tau^{-\frac{2}{3}}$ is the same as in the first phase. It is interesting to note that its validity extends considerably into the second phase up to a point $q = 2{\cdot}45$. In the first phase the heat does not penetrate into the fluid beyond a distance δ corresponding to $q = 1$, while at the point $q = 2{\cdot}45$ the heat has penetrated to a distance $2{\cdot}45\delta$.

The value $0{\cdot}554\tau^{-\frac{1}{2}}$ is the asymptotic value for large q and is the same as would be obtained by the variational method for a constant velocity profile $\varphi(\eta) = 1$ throughout.

The point $\tau = 0{\cdot}64$ is one for which the two analytical expressions $0{\cdot}514\tau^{-\frac{2}{3}}$ and $0{\cdot}554\tau^{-\frac{1}{2}}$ are equal.

As can be seen from Table 7.1, the approximation (5.9) is quite accurate except for a deviation that is not significant in a restricted range near the point $\tau = 0{\cdot}64$.

Comparison with exact solutions

In the first phase the variational solution is the same as for the case where the linear velocity profile extends to infinity. Hence in this case

$$\varphi(\eta) = \eta \quad \text{for } 0 < \eta < \infty. \tag{5.9 a}$$

The problem becomes one of solving the equation

$$\eta\frac{\partial\theta}{\partial\tau} = \frac{\partial^2\theta}{\partial\eta^2}, \tag{5.9 b}$$

valid for the complete range of positive values of η. It is readily verified that an exact solution of equation (5.9 b) is

$$\theta = C\tau^{-\frac{2}{3}}\exp\left(-\frac{\eta^3}{9\tau}\right) \tag{5.9 c}$$

with a constant factor C. This solution satisfies the boundary condition $\partial\theta/\partial\eta = 0$ at $\eta = 0$. The constant C is determined by the integral condition (4.11). We find

$$C = \frac{H_0}{3^{\frac{1}{3}}\Gamma(\frac{2}{3})} \tag{5.9 d}$$

where Γ denotes the gamma function. The value θ_0 of θ for $\eta = 0$ is

$$\theta_0 = 0{\cdot}512 H_0 \tau^{-\frac{2}{3}}. \tag{5.9 e}$$

Hence the reduced trailing function is

$$\phi(\tau) = \frac{\theta_0}{H_0} = 0{\cdot}512\tau^{-\frac{2}{3}}. \tag{5.9 f}$$

The approximate expression (5.7) derived by the variational procedure is

$$\phi(\tau) = 0{\cdot}514\tau^{-\frac{2}{3}}. \tag{5.9 g}$$

Comparing with the exact value (5.9 f) we see that the error is less than one-half of 1 per cent.

At the other extreme we consider the asymptotic solution of equations (5.3) for large values of τ. This corresponds to the case where the velocity profile is a constant $u = U$, hence $\varphi(\eta) = 1$. The exact solution for this case was already derived in Section 3 and is expressed by equation (3.24) which may be written

$$\theta_0 = \frac{1}{\sqrt{(\pi c U k x)}} = \frac{H_0}{\sqrt{(\pi\tau)}}. \tag{5.9 h}$$

Hence

$$\phi(\tau) = 0{\cdot}565\tau^{-\frac{1}{2}}. \tag{5.9 i}$$

Comparing this exact value with the approximate solution $0{\cdot}554\tau^{-\frac{1}{2}}$ of equation (4.10), derived by the variational method for $\varphi(\eta) = 1$ using the approximation (4.13) we see that the error is about 2 per cent. Note that this is larger than the error obtained in Section 3 by using the parabolic approximation (3.13) for the temperature. The larger error is due to the use of the cubic (4.13) for the temperature instead of the parabola. Preference has been given to the cubic approximation because it yields a higher accuracy in the first phase which is the more significant one.

Parabolic velocity profile

Consider next a boundary layer with the velocity profile shown in Fig. 7.2 (*b*). The velocity profile is parabolic within a thickness 2δ beyond which the velocity is constant and equal to U. The function $\varphi(\eta)$ corresponding to this profile is

$$\begin{aligned} \varphi(\eta) &= \eta(1-\tfrac{1}{4}\eta) \quad \text{for } \eta < 2, \\ \varphi(\eta) &= 1 \quad \text{for } \eta > 2. \end{aligned} \tag{5.10}$$

The characteristic thickness δ in this case is the same as in the piece-wise linear profile obtained by drawing a tangent to the parabola at $y = 0$.

The trailing function for this parabolic profile is obtained by inserting the value (5.10) for $\varphi(\eta)$ into the general equations of Section 4. The result has been derived numerically.‡ For most purposes it is found

‡ See the author's paper cited on p. 117, n. §.

that the reduced trailing function $\phi(\tau)$ may be approximated by the piece-wise analytical expressions (5.9). The approximation is good in the significant range where $\tau < 0{\cdot}1$. However, in the region near $\tau = 0{\cdot}6$ the error is of the order of 10 per cent and decreases again for larger values of τ. An improved approximation is obtained by adding a small analytical term as follows:

$$\begin{aligned} \phi(\tau) &= 0{\cdot}514\tau^{-\frac{2}{3}}+\frac{0{\cdot}66\tau}{1+9\tau^2} \quad \text{for } \tau < 0{\cdot}64, \\ \phi(\tau) &= 0{\cdot}554\tau^{-\frac{1}{2}}+\frac{0{\cdot}66\tau}{1+9\tau^2} \quad \text{for } \tau > 0{\cdot}64. \end{aligned} \tag{5.11}$$

This reduces the maximum error to a value of the order of 2 per cent.

We may conclude from this analysis that the trailing function is not sensitive to the velocity profile, provided the characteristic thickness δ is chosen as defined in Fig. 7.2 by drawing a tangent to the velocity profile at the wall. The value δ is the distance from the wall of the point of intersection of this tangent with the line $u = U$ representing the constant velocity outside the boundary layer. For a family of velocity profiles between the two extreme cases examined here, when high accuracy is not required, the approximate value (5.9) of $\phi(\tau)$ will be adequate.

6. TURBULENT BOUNDARY LAYER

We shall now adapt the general procedure of Section 4 to derive an approximate expression of the trailing function for a turbulent boundary layer. We shall consider the case of parallel streamlines.

In the original analysis of the turbulent boundary layer by von Kármán‡ and later authors it is possible to introduce a characteristic thickness

$$\delta = 14\nu\sqrt{\left(\frac{\rho}{S}\right)}, \tag{6.1}$$

where S is the shear stress at the wall, ρ the mass density of the fluid, and ν its kinematic viscosity. (The viscosity coefficient is equal to $\nu\rho$.)

The velocity profile of the turbulent boundary layer is shown in Fig. 7.3. There are three distinct regions. They are:

(*a*) the laminar sublayer for

$$y < \tfrac{1}{3}\delta;$$

‡ Th. von Kármán, 'Some aspects of the turbulence problem', *Proc. fourth Int. Congr. Appl. Mech.* (1934), pp. 54–91. Cambridge University Press (1935). Also by the same author, 'The analogy between fluid friction and heat transfer', *Trans. Am. Soc. mech. Engrs.* **61**, 705–10 (1939).

(*b*) the buffer layer for

$$\tfrac{1}{3}\delta < y < 2\delta;$$

(*c*) the fully turbulent region for

$$y > 2\delta.$$

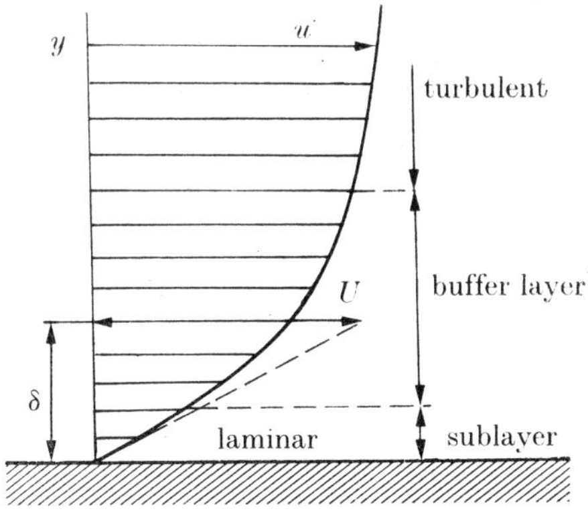

FIG. 7.3. Velocity profile for the turbulent boundary layer. Reference thickness δ and reference velocity *U*.

It is convenient here to adopt as reference velocity the expression

$$U = 14\sqrt{\left(\frac{S}{\rho}\right)}. \tag{6.2}$$

As illustrated in Fig. 7.3, this velocity is the same as would exist at the distance $y = \delta$ if the velocity were linearly distributed and obtained by drawing a tangent to the profile at the wall. This can be seen by combining equations (6.1) and (6.2) and writing

$$S = \nu\rho\frac{U}{\delta}. \tag{6.3}$$

The Reynolds number corresponding to U and δ is

$$\frac{U\delta}{\nu} = 196. \tag{6.4}$$

In many problems of convective heat transfer, use is made of the Prandtl number, defined as

$$Pr = \frac{k}{\nu c}. \tag{6.5}$$

Combining relations (6.4) and (6.5) we may write the Peclet number (4.9) as

$$Pe = \frac{196}{Pr}. \tag{6.6}$$

The present description of the turbulent boundary layer is, of course, approximate. However, for our purpose it may be assumed that it does not deviate significantly from reality except in extreme cases.

It is possible to express the velocity profile of the turbulent boundary layer in the form

$$u = U\varphi(\eta), \qquad \eta = \frac{y}{\delta}, \tag{6.7}$$

where δ and U are defined by equations (6.1) and (6.2). In order to apply the general procedure of Section 4, we must integrate equation (4.36). This requires the knowledge of the function $\sigma(\eta)$, which depends on the distribution of molecular and turbulent diffusivity in the boundary layer. For this purpose we may adopt a representation of this function proposed by Rannie‡ as follows:

$$\sigma(\eta) = 1+\sinh^2\eta \quad \text{for } \eta < 2, \qquad \sigma(\eta) = 6{\cdot}6\eta \quad \text{for } \eta > 2. \tag{6.8}$$

This representation assumes that the Prandtl number is not far from unity. The important point here is the general behaviour of the function $\sigma(\eta)$, since the results derived below are not very sensitive to detailed inaccuracies of the representation. Note that the value of η for which the two values (6.8) of $\sigma(\eta)$ are equal, is slightly smaller than 2. Using a standard approximation for the turbulent velocity profile $\varphi(\eta)$ and the approximate value (6.8) of $\sigma(\eta)$, we derive the variable η' and the function $\beta(\eta')$ given by equations (4.35) and (4.37) of the general theory. A typical plot of the function $\beta(\eta')$ is shown in Fig. 7.4. The curve may generally be approximated by two straight lines:

$$\beta(\eta') = \eta' \quad \text{for } \eta' < 1, \qquad \beta(\eta') = 1 \text{ to } \infty \quad \text{for } \eta' = 1. \tag{6.9}$$

Since $\eta' \simeq \eta$ for $\eta' < 1$, this approximation may be replaced by

$$\beta(\eta) = \eta \quad \text{for } \eta < 1, \qquad \beta(\eta) = 1 \text{ to } \infty \quad \text{for } \eta = 1. \tag{6.10}$$

Hence in the evaluation of the trailing function we may again distinguish two phases depending on whether η is smaller or larger than 1.

In the first phase, where $\eta < 1$, the differential equation (4.36) is reduced to

$$\eta\frac{\partial\theta}{\partial\tau} = \frac{\partial^2\theta}{\partial\eta^2}. \tag{6.11}$$

This coincides with the first of equations (5.3). Hence in the first phase the trailing function is the same as given by equation (5.7) for the

‡ W. D. Rannie, 'Heat transfer in turbulent shear flow', *J. aeronaut. Sci.* **23**, 485–9 (1956).

laminar boundary layer with a linear velocity profile. This reduced trailing function is

$$\frac{\theta_0}{H_0} = \phi(\tau) = 0{\cdot}514\tau^{-\frac{2}{3}}. \tag{6.12}$$

At the end of this first phase the penetration depth is $q = \eta = 1$. According to Table 7.1 the corresponding value of τ is

$$\tau_t = 0{\cdot}0606. \tag{6.13}$$

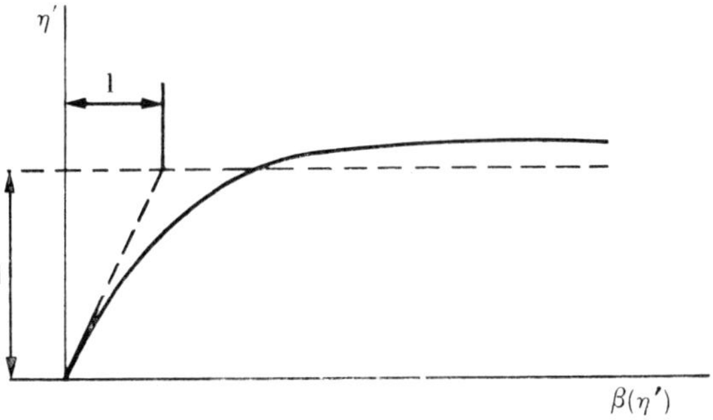

FIG. 7.4. Plot of $\beta(\eta')$. Dotted line shows the piece-wise linear approximation (6.10).

The second phase starts for $\tau > \tau_t$. This is the point at which the effect of the turbulent diffusivity begins to enter into the picture. We have therefore referred to the point $\tau = \tau_t$ as the *turbulent cross-over point.*

In the second phase we must solve the equation

$$\beta(\eta)\frac{\partial\theta}{\partial\tau} = \frac{\partial^2\theta}{\partial\eta^2} \tag{6.14}$$

with the values (6.10) for $\beta(\eta)$. As can be seen, this amounts to solving equation (6.11) with the boundary condition

$$\theta = 0 \quad \text{at } \eta = 1. \tag{6.15}$$

The temperature distribution in the region $\eta < 1$ at the end of the first phase and during the second phase is obtained by putting $q = 1$ in equation (4.13). Hence

$$\theta = \theta_0(1-\eta^3). \tag{6.16}$$

This expression satisfies the required boundary condition (6.15). The value of θ_0 is a function of τ which we shall now derive by applying the variational method to equation (6.14), using the cubic approximation (6.16). From the standpoint of the conduction analogy the second phase represents a leakage of heat from the region $\eta < 1$ into an adjacent medium of infinite specific heat.

With the value (6.16) for θ and $\beta(\eta) = \eta$, the thermal potential corresponding to equation (6.14) is

$$V = \tfrac{1}{2}\int\limits_0^1 \eta\theta^2\,d\eta = \tfrac{9}{80}\theta_0^2. \tag{6.17}$$

Because of the adiabatic condition $\dot{H} = 0$ for $\eta = 0$, we may write

$$\dot{H} = -\int\limits_0^\eta \dot{\theta}\eta\,d\eta = -(\tfrac{1}{2}\eta^2 - \tfrac{1}{5}\eta^5)\dot{\theta}_0. \tag{6.18}$$

Hence the dissipation function is

$$D = \tfrac{1}{2}\int\limits_0^1 \dot{H}^2\,d\eta = \tfrac{63}{4400}\dot{\theta}_0^2. \tag{6.19}$$

The dot symbol represents the derivative with respect to τ. The unknown generalized coordinate is the variable θ_0. It satisfies the Lagrangian equation

$$\frac{\partial V}{\partial \theta_0} + \frac{\partial D}{\partial \dot{\theta}_0} = 0. \tag{6.20}$$

With expressions (6.17) and (6.19) this yields the differential equation

$$55\theta_0 + 7\dot{\theta}_0 = 0. \tag{6.21}$$

This must be solved with the initial condition that θ_0 is equal to its value at the end of the first phase. This value obtained by substituting $\tau = \tau_t$ in expression (6.12) is

$$\theta_0 = 3{\cdot}33H_0. \tag{6.22}$$

The corresponding solution of equation (6.21) is

$$\theta_0 = 3{\cdot}33H_0 \exp\{-\tfrac{55}{7}(\tau - \tau_t)\}. \tag{6.23}$$

From equations (6.12) and (6.23) we conclude that the reduced trailing function for both the first and second phase may be represented by the piece-wise analytical approximation

$$\begin{aligned} \phi(\tau) &= 0{\cdot}514\tau^{-\frac{3}{8}} && \text{for } \tau < \tau_t, \\ \phi(\tau) &= 3{\cdot}33\exp\{-\tfrac{55}{7}(\tau - \tau_t)\} && \text{for } \tau > \tau_t, \end{aligned} \tag{6.24}$$

where $\tau_t = 0{\cdot}0606$. This reduced trailing function is plotted in Fig. 7·5. In contrast with the trailing function (5.9) for laminar flow, which is also plotted for comparison, it decays much faster from the point of injection. In fact beyond the point $\tau = 0{\cdot}6$ it may be considered as having vanished and we may put $\phi(\tau) = 0$.

The approximate trailing function (6.24), which has been derived here by a partial modification of the general procedure of Section 4, is based on a typical standard physical description of the turbulent boundary layer.

It should provide a satisfactory first approximation for the evaluation of convective heat transfer for turbulent boundary layers in many practical applications. Refinements in accuracy may be obtained if needed by following procedures similar to those described in the foregoing analysis.

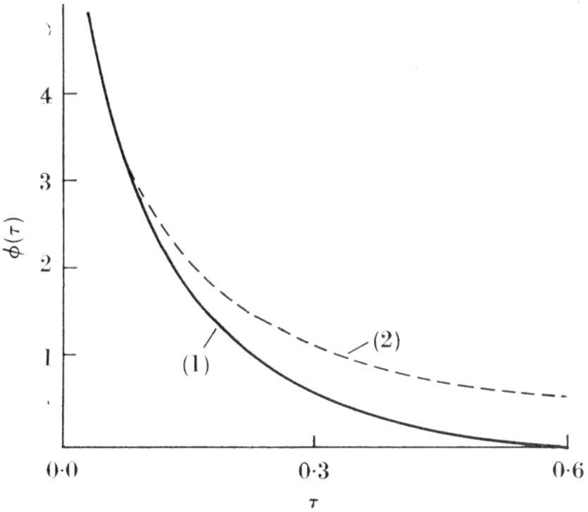

FIG. 7.5. Reduced trailing function $\phi(\tau)$. (1) For the turbulent boundary layer as given by equations (6.24). (2) For the laminar boundary layer as given by equations (5.9).

7. APPLICATIONS

We shall illustrate here the formulation of typical problems of convective heat transfer by considering first the free stream boundary. It will then be shown how a slight modification of the trailing function leads to a general formulation of heat transfer for ducted flow and heat exchangers.

Free stream boundary layer

In this case we must generally take into account the fact that the streamlines are non-parallel. The trailing function in this case will depend on the abscissa ξ of the point of injection. As pointed out at the end of Section 4, under certain conditions we may use the function $\phi(\tau)$ evaluated above for parallel streamlines as a first approximation for non parallel streamlines. This procedure will be valid if the boundary layer does not vary too rapidly streamwise and if the significant part of the heat transfer occurs in a relatively small distance down-stream from the point of injection. The trailing function may then be written as

$$r(x-\xi,\xi) = \frac{1}{kPe(\xi)}\phi(\tau), \tag{7.1}$$

with

$$\tau = \frac{x-\xi}{Pe(\xi)\delta(\xi)}. \tag{7.2}$$

This is derived from equation (4.30) replacing x by the down-stream distance $x-\xi$ from the point of injection. We must introduce here a local reference thickness $\delta(\xi)$ and a local reference velocity $U(\xi)$. Both $\delta(\xi)$ and $U(\xi)$ depend on the point of injection ξ, while

$$Pe(\xi) = \frac{c}{k}U(\xi)\delta(\xi) \tag{7.3}$$

is a local Peclet number.

This procedure is applicable to either laminar or turbulent flow. Note that for turbulent flow the value of $Pe(\xi)$ is given by equation (6.6) in terms of the Prandtl number, which is independent of ξ.

A more accurate value of the trailing function (7.1) for non-parallel streamlines could be obtained by applying the variational method to the conduction analogy represented by equation (2.17).

According to the definition of the one-dimensional trailing function in Section 2 of Chapter 6, we may express the excess temperature θ at the wall over the adiabatic temperature θ_a as

$$\theta-\theta_a = \int^{\xi} \dot{H}(\xi,t)r(x-\xi,\xi)\,d\xi, \tag{7.4}$$

where $\dot{H}$ is the rate of injection of heat into the fluid at the wall per unit area at point ξ and time t. Expression (7.4) assumes a steady velocity field. If the fluid flow is time-dependent the trailing function r will also depend on t.

Equation (7.4) yields the wall temperature directly when the distribution of heat injection $\dot{H}$ into the fluid is given. This corresponds to forced convection. When the wall temperature $\theta-\theta_a$ is given, equation (7.4) is an integral equation for the unknown rate of heat transfer $\dot{H}$. It may be solved numerically by standard programming techniques. In particular, we may choose a polynomial representation of $\dot{H}$ as a function of ξ with unknown coefficients; and equate the values of each side of equation (7.4) at given points x.

Trailing function for ducted flow

We consider the case of a two-dimensional flow field between parallel walls separated by a distance D and a location down-stream uninfluenced by entrance conditions into the duct. Hence we may assume parallel streamlines, and the trailing function is expressed as $r(x)$, where x is the distance down-stream from the point of injection. It differs from the trailing function derived above for the free-stream boundary by the fact that for large values of x it tends toward a non-vanishing asymptotic value. This is due to the fact that the injected heat at large distance down-stream is diluted uniformly into a finite volume of fluid. This asymptotic value is

$$r(\infty) = \frac{1}{cDu_{\text{av}}}. \tag{7.5}$$

The average flow velocity u_{av} is defined as the rate of volume flow divided by the cross-section. The asymptotic value of the reduced trailing function is derived from equation (4.3). We obtain

$$\phi(\infty) = Pe\,kr(\infty) = \frac{\delta U}{Du_{\text{av}}}. \tag{7.6}$$

Problems of heat transfer for ducted flows may therefore be treated by slight modification of the trailing function, taking into account the asymptotic value (7.6). A simple procedure is to add the asymptotic value to the function $\phi(\tau)$ derived in sections 5 and 6.

When entrance conditions must be taken into account, the trailing function is expressed as $r(x-\xi, \xi)$. It depends on the point of injection and has the asymptotic value (7.5).

Heat exchangers

Consider a two-dimensional flow of two fluids flowing in opposite directions and separated by a thin solid wall. We assume steady flow and time-independent conditions. The first fluid flows in the x direction. Its trailing function is $r(x-\xi, \xi)$ and its adiabatic temperature is $\theta_a(x)$. The heat transfer occurs between the abscissae x_0 and x_1. The wall temperature for the first fluid is

$$\theta(x) = \theta_a(x) + \int\limits_{x_0}^{x_1} r(x-\xi, \xi)\dot{H}(\xi)\,d\xi, \tag{7.7}$$

where $\dot{H}(\xi)$ is the local rate of heat transfer from the second to the first fluid. Since we are dealing with ducted flows, the trailing function $r(x-\xi, \xi)$ has an asymptotic value as indicated in the foregoing paragraph. Entrance conditions may be taken into account.

The second fluid flows in the negative direction of x. Its trailing function is $r'(\xi-x, \xi)$ and its adiabatic temperature is $\theta'_a(x)$. The wall temperature of the second fluid is

$$\theta'(x) = \theta'_a(x) - \int_{x_0}^{x_1} r'(\xi - x, \xi)\dot{H}(\xi)\, d\xi. \tag{7.8}$$

The temperature difference across the separating wall is

$$\theta'(x) - \theta(x) = \frac{a}{k_s}\dot{H}(x), \tag{7.9}$$

where a is the thickness of the separating wall and k_s its thermal conductivity. By combining relations (7.7), (7.8), and (7.9) we derive

$$\theta'_a(x) - \theta_a(x) = \int_{x_0}^{x_1} R(x, \xi)\dot{H}(\xi)\, d\xi + \frac{a}{k_s}\dot{H}(x) \tag{7.10}$$

with

$$R(x, \xi) = r'(\xi - x, \xi) + r(x - \xi, \xi). \tag{7.11}$$

This is an integral equation for the unknown heat transfer distribution $\dot{H}(x)$. It may be solved numerically by standard programming procedures. In most cases the adiabatic temperature $\theta'_a - \theta_a$ will be independent of x.

For fully-developed duct flow independent of entrance conditions we may assume parallel streamlines, and expression (7.11) becomes

$$R(x - \xi) = r'(\xi - x) + r(x - \xi), \tag{7.12}$$

where $r'(\xi - x)$ and $r(x - \xi)$ are the trailing functions of the two fluids.

CHAPTER EIGHT

COMPLEMENTARY PRINCIPLES

1. INTRODUCTION

In the mechanics of deformable systems variational principles may be expressed in dual form. In one of these, the variables subject to variations are displacements, which are chosen so as to satisfy the condition of continuity. The variational principle formulated by means of these displacements then expresses the laws of equilibrium of the forces. In another form, known as *complementary*, the forces are the variables subject to variation and are chosen to satisfy equilibrium conditions. The complementary variational principle formulated in terms of these forces then expresses continuity of displacement.

In the preceding chapters we have considered variational principles in heat transfer which we have referred to as 'fundamental'. They are expressed by means of variations applied to the heat displacement chosen so as to satisfy conservation of energy. This amounts to satisfying a continuity equation as a constraint and expressing the response of heat displacement to temperature in variational form.

The analogy with mechanics suggests the existence of a complementary form of the variational principles in heat transfer. Since the temperature plays the role of a force, we must vary the temperature field and express in variational form the condition of continuity of heat flow, which in this case is also the condition of energy conservation. In the following discussion of these complementary principles we shall follow closely the material contained in the author's paper.‡

In Section 2 the complementary principles are discussed in the context of linear systems. It is shown that their formulation in terms of generalized coordinates leads to equations of the Lagrangian type in complementary form. An interconnection principle applicable to a 'finite element' method may also be formulated and, as pointed out, retains its validity in the general case of non-linearity and convection. In Section 3 we discuss an operational formulation of the variational principle complementary to the results obtained in Chapter 3. Non-

‡ M. A. Biot, 'Complementary forms of the variational principle for heat conduction and convection', *J. Franklin Inst.* **283**, 372–8 (1967).

linear systems with temperature-dependent properties both stationary and convective are considered in Sections 4 and 5.

A remark must be made here regarding the accuracy obtained by applying complementary principles. The latter involve a space differentiation of the temperature field. As a consequence, when using the same approximate representation of the temperature, complementary principles will generally yield results that are less accurate than those formulated in terms of the fundamental form based on heat displacements.

It should also be noted that the existence of dual complementary forms of variational principles is a very general property of physical systems that are described by intensive and extensive variables. The method is therefore susceptible to wide generalizations in many domains of physics.

2. CONDUCTION IN LINEAR SYSTEMS

We consider thermal conduction in a medium with properties independent of the temperature θ. The heat capacity per unit volume is

$$c = c(x, y, z). \tag{2.1}$$

It may be a function of the coordinates. Assuming isotropic properties, the thermal conductivity

$$k = k(x, y, z, t) \tag{2.2}$$

may be a function of time and the coordinates.

The law of heat conduction is

$$\mathbf{J} = -k \operatorname{grad} \theta \tag{2.3}$$

and conservation of energy is expressed as

$$c\dot{\theta} = -\operatorname{div} \mathbf{J}. \tag{2.4}$$

The procedure used in Section 2 of Chapter 1 amounted to satisfying identically the equation of energy conservation (2.4) by putting

$$\begin{aligned} c\theta &= -\operatorname{div} \mathbf{H}, \\ J &= \dot{\mathbf{H}}. \end{aligned} \tag{2.5}$$

The law of thermal conduction (2.3) is then satisfied approximately by a variational principle expressed in terms of the variations $\delta\mathbf{H}$.

This suggests a complementary procedure where $\mathbf{J}$ is defined by equation (2.3) in terms of a scalar field θ. The equation of energy conservation (2.4) is then satisfied approximately by a variational principle expressed in terms of the variation $\delta\theta$ of the temperature field.

We therefore multiply equation (2.4) by $\delta\theta$ and integrate the result over the volume τ of the thermal system. We obtain

$$\iiint_\tau (c\dot{\theta}+\operatorname{div}\mathbf{J})\,\delta\theta\,d\tau = 0. \tag{2.6}$$

With arbitrary variations $\delta\theta$ this is equivalent to equation (2.4) in the context of weak solutions. Integration by parts of the integral (2.6) yields

$$\iiint_\tau c\dot{\theta}\,\delta\theta\,d\tau - \iiint_\tau \mathbf{J}.\operatorname{grad}\delta\theta\,d\tau + \iint_A \mathbf{J}.\mathbf{n}\,\delta\theta\,dA = 0. \tag{2.7}$$

The surface integral is extended to boundary A of the volume τ and $\mathbf{n}$ denotes the unit outward normal to the boundary. Since $\mathbf{J}$ is defined by equation (2.3) we may write

$$-\iiint_\tau \mathbf{J}.\operatorname{grad}\delta\theta\,d\tau = \delta D, \tag{2.8}$$

where

$$D = \tfrac{1}{2}\iiint_\tau k(\operatorname{grad}\theta)^2\,d\tau \tag{2.9}$$

is physically identical with the dissipation function (4.3) of Chapter 1. By taking into account relation (2.8), equation (2.7) becomes

$$\iiint_\tau c\dot{\theta}\,\delta\theta\,d\tau + \delta D = -\iint_A \mathbf{J}.\mathbf{n}\,\delta\theta\,dA. \tag{2.10}$$

This is the complementary form of the variational principle for thermal conduction.

An important remark must be made here regarding the significance of the variational equation (2.10). The value of $\mathbf{J}$ on the right side may be put equal to the value of $-k\operatorname{grad}\theta$ at the boundary. The heat flux $\mathbf{J}.\mathbf{n}$ at the boundary may also be considered as a given function of time and location. In this case the variational principle (2.10) acquires an additional significance. This is brought out by integrating δD by parts. We write

$$\delta D = -\iiint_\tau \operatorname{div}(k\operatorname{grad}\theta)\,\delta\theta\,d\tau + \iint_A (k\operatorname{grad}\theta).\mathbf{n}\,\delta\theta\,dA. \tag{2.11}$$

With this value equation (2.10) becomes

$$\iiint_\tau \{c\dot{\theta}-\operatorname{div}(k\operatorname{grad}\theta)\}\,\delta\theta\,d\tau = -\iint_A \{\mathbf{J}.\mathbf{n}+k(\operatorname{grad}\theta).\mathbf{n}\}\,\delta\theta\,dA. \tag{2.12}$$

Since $\delta\theta$ is arbitrary, this equation implies that the equation

$$c\dot{\theta} = \operatorname{div}(k\operatorname{grad}\theta) \tag{2.13}$$

is verified in the volume τ and, in addition, that the condition

$$\mathbf{J}.\mathbf{n} = -k(\operatorname{grad}\theta).\mathbf{n} \tag{2.14}$$

is verified at the boundary.

Hence, when the boundary heat flux $\mathbf{J}.\mathbf{n}$ is prescribed, the complementary variational principle (2.10) provides a method of satisfying approximately the boundary condition (2.14) and the classical equation (2.13) for thermal diffusion. As can be seen, equation (2.13) expresses conservation of energy, since it coincides with relation (2.4) after substitution of the value (2.3) for $\mathbf{J}$.

A variational principle that bears some analogy to the variational principle (2.10) has also been proposed by Chambers.‡ It is obtained by considering the functional

$$F = \tfrac{1}{2}\iiint_{\tau} \{cX^2+k(\operatorname{grad}\theta)(\operatorname{grad}X)-wX\}\,d\tau+\iint_{A} \mathbf{J}.\mathbf{n}X\,dA, \tag{2.14 a}$$

where $X = \dot{\theta}$ and w is the heat generated per unit time and unit volume. The variable to be varied is X. The variational principle is expressed by putting $\delta F = 0$ for all arbitrary variations of X.

Another method proposed by Rosen§ is obtained by considering the functional

$$\phi = \iiint_{\tau} \{c\theta Y+\tfrac{1}{2}k(\operatorname{grad}\theta)^2\}\,d\tau. \tag{2.14 b}$$

The variable θ is varied inside the volume τ, while $Y = \dot{\theta}$ is not subject to variation. With these restrictions the variational principle is expressed by $\delta\phi = 0$.

A modification of this method follows a suggestion by Prigogine and Glansdorff adapted to heat transfer by Hays.‖ The variational principle is obtained by putting $\delta\Psi = 0$ where Ψ is the time integral of ϕ

$$\Psi = \int_{t_0}^{t_1} \phi\,dt. \tag{2.14 c}$$

The value of θ is varied, assuming that it is a given function of time and the coordinates and of a certain number of adjustable constant parameters that are evaluated by applying the variational principle. The procedure is applicable to non-linear problems where k and c are temperature-dependent, provided these quantities are not subject to variation. Note that if θ is chosen to be a linear function of the parameters this procedure coincides with the Galerkin method as can be seen from the discussion in Section 4 of the Appendix.

By following a procedure similar to that of Chapter 1, the variational principle (2.10) also leads to differential equations analogous to

‡ L. G. Chambers, 'A variational principle for the conduction of heat', *Q. Jl Mech. appl. Math.* **9**, 234–5 (1956).

§ P. Rosen, 'On variational principles for irreversible processes', *J. Chem. Phys.* **21**, 1220–1 (1953).

‖ See, for example, D. F. Hays and H. N. Curd, 'Heat conduction in solids: temperature-dependent thermal conductivity', *Int. J. Heat Mass Transfer*, **11**, 285–95 (1968).

Lagrangian equations with generalized coordinates. This can be shown by expressing the unknown temperature field as

$$\theta = \theta(q_1, q_2, \ldots, q_n, x, y, z, t), \tag{2.15}$$

where the q_i are n unknown functions of time representing generalized coordinates.

As already pointed out in Section 3 of Chapter 1, the introduction of generalized coordinates does not restrict the generality of the analysis. Because of the discontinuous nature of matter, a finite, but sufficiently large, number of generalized coordinates can always be found to describe the field θ with adequate physical accuracy.‡

The variations $\delta\theta$ depend only on the variations δq_i. Hence expression (2.15) yields

$$\delta\theta = \sum^{i} \frac{\partial\theta}{\partial q_i}\,\delta q_i. \tag{2.16}$$

From expression (2.15) we also derive

$$\dot\theta = \sum^{i} \frac{\partial\theta}{\partial q_i}\dot q_i + \frac{\partial\theta}{\partial t}. \tag{2.17}$$

Note that in this equation $\dot\theta$ is the total time derivative of θ at a fixed point x, y, z. Equation (2.17) yields

$$\frac{\partial\dot\theta}{\partial\dot q_i} = \frac{\partial\theta}{\partial q_i}. \tag{2.18}$$

As a consequence, the variation (2.16) may also be written

$$\delta\theta = \sum^{i} \frac{\partial\dot\theta}{\partial\dot q_i}\,\delta q_i. \tag{2.19}$$

Finally, the variation of D is

$$\delta D = \sum^{i} \frac{\partial D}{\partial q_i}\,\delta q_i. \tag{2.20}$$

By substituting the values (2.16), (2.19), and (2.20) into the variational equation (2.10), we derive

$$\sum^{i} \delta q_i\left(\iiint_{\tau} c\theta\frac{\partial\dot\theta}{\partial\dot q_i}\,d\tau + \frac{\partial D}{\partial q_i}\right) = -\sum^{i} \iint_{A} \mathbf{J}\,.\,\mathbf{n}\frac{\partial\theta}{\partial q_i}\,dA\,\delta q_i. \tag{2.21}$$

Since δq_i is arbitrary, this leads to n equations:

$$\frac{\partial G}{\partial\dot q_i} + \frac{\partial D}{\partial q_i} = P_i, \tag{2.22}$$

‡ This is also discussed in Section 4 of the Appendix.

where
$$G = \tfrac{1}{2}\iiint\limits_{\tau} c\dot{\theta}^2\, d\tau,$$
$$P_i = -\iint\limits_{A} \mathbf{J}.\mathbf{n}\frac{\partial\theta}{\partial q_i}\, dA. \tag{2.23}$$

The n differential equations (2.22) for the n unknown function of time q_i may be considered as a complementary form of the Lagrangian equations.

Let us consider the case of a linear representation of the temperature field in terms of generalized coordinates. We write‡
$$\theta = \sum^{i} q_i\,\theta_i(x, y, z), \tag{2.24}$$
where θ_i are given time-independent scalar fields. In this case the values of G and D, as defined by equations (2.23) and (2.9), are quadratic forms
$$G = \tfrac{1}{2}\sum^{ij} a_{ij}\,\dot{q}_i\,\dot{q}_j,$$
$$D = \tfrac{1}{2}\sum^{ij} b_{ij}\,q_i\,q_j, \tag{2.25}$$
where
$$a_{ij} = \iiint\limits_{\tau} c\theta_i\,\theta_j\, d\tau,$$
$$b_{ij} = \iiint\limits_{\tau} k(\operatorname{grad}\theta_i)(\operatorname{grad}\theta_j)\, d\tau. \tag{2.26}$$
Also
$$P_i = -\iint\limits_{A} \mathbf{J}.\mathbf{n}\,\theta_i\, dA. \tag{2.27}$$

The Lagrangian equations (2.22) become
$$\sum^{j} a_{ij}\,\dot{q}_j + \sum^{j} b_{ij}\,q_j = P_i. \tag{2.28}$$
They are the complementary form of the linear equations (3.14) of Chapter 2. When the heat flux $\mathbf{J}$ is prescribed at the boundary, the values of P_i are known functions of time which behave as 'driving forces'.

The variational principle (2.10) is readily extended to anisotropic conductivity
$$k_{ij} = k_{ji} = k_{ij}(x, y, z, t), \tag{2.29}$$
including the case where heat is generated in the body at a rate
$$w = w(x, y, z, t) \tag{2.30}$$
per unit volume. The variational principle (2.10) is replaced by
$$\iiint\limits_{\tau} (c\dot{\theta} - w)\,\delta\theta\, d\tau + \delta D = -\iint\limits_{A} \mathbf{J}.\mathbf{n}\,\delta\theta\, dA, \tag{2.31}$$

‡ Note that if the temperature is prescribed at the boundary we may write $\theta = \sum^{i} q_i\,\theta_i + \theta_b$, where θ_b is a function of the coordinates and the time assuming the prescribed boundary values, while $\theta_i = 0$ at the boundary. The values of G and D are then second degree polynomials in $\dot{q}_i$ and q_i respectively and $P_i = 0$.

where D is now defined as

$$D = \tfrac{1}{2} \iiint \sum^{ij} k_{ij} \frac{\partial\theta}{\partial x_i} \frac{\partial\theta}{\partial x_j}\, d\tau. \tag{2.32}$$

As before, the coordinates x, y, z are represented by x_i. With this expression of D, the Lagrangian equations (2.22) remain valid. In addition, if there are heat sources we must replace the value (2.27) of P_k by

$$P_k = -\iint_A \mathbf{J}\,.\,\mathbf{n} \frac{\partial\theta}{\partial q_k}\, dA + \iiint_\tau w \frac{\partial\theta}{\partial q_k}\, d\tau. \tag{2.33}$$

When the thermal sources and the boundary heat flux are prescribed, the quantities P_k play the role of driving forces in the Lagrangian equations.

Interconnection and finite element method

In Section 7 of Chapter 3 we have discussed an interconnection principle whereby a system may be divided into a number of finite elements which are then treated as interconnected sub-systems. The same procedure is applicable to the complementary form of the variational principle.‡

For simplicity, and without loss of generality, we may consider the case of isotropic conductivity without thermal sources. We divide the system into a number of interconnected domains denoted by a number s. In each domain s the temperature field is approximated by θ_s in such a way that these temperatures coincide at adjacent boundaries. The variational principle (2.10) for the total system may be written as a sum of equations for each domain:

$$\sum^{s} \iiint_{\tau_s} c\dot{\theta}_s\, \delta\theta_s\, d\tau_s + \sum^{s} \delta D_s = -\sum^{s} \iint_{A_s} \mathbf{J}_s\,.\,\mathbf{n}_s\, \delta\theta_s\, dA_s. \tag{2.34}$$

The basic property of interconnection results from the expression on the right side of this equation. Common boundaries appear twice in the summation and the corresponding terms cancel each other. This can be seen by considering two adjacent sub-systems, s and $s+1$. At the common boundary the normal heat fluxes due to J_s and J_{s+1} are the same, as well as the temperatures θ_s and θ_{s+1}. However, the normal directions $\mathbf{n}_s$ and $\mathbf{n}_{s+1}$ are opposite ($\mathbf{n}_s = -\mathbf{n}_{s+1}$). Hence the terms corresponding

‡ Note the difference of this formulation from another complementary principle derived for finite elements and linear systems in Section 7 of Chapter 3.

to adjacent boundaries cancel out and we may write

$$\sum^{s} \iint_{A_s} \mathbf{J}.\mathbf{n}_s\, \delta\theta_s\, dA_s = \iint_{B} \mathbf{J}.\mathbf{n}\, \delta\theta\, dB, \tag{2.35}$$

where B designates the outer boundary of the total system.

In applying the variational principle (2.34) we may separate the generalized coordinates into two groups, one defining the temperatures at the boundaries of the sub-system and the other defining the temperature in each sub-system. This leads to Lagrangian equations which involve the heat flux $\mathbf{J}$ only at the outer boundary B of the total system. We note that in this procedure it is not necessary to impose a condition of continuity of $k \operatorname{grad} \theta$ in a direction normal to interconnecting boundaries. In a two-dimensional problem, for example, the system could be divided into finite elements by a triangular network, and some of the generalized coordinates would be the temperatures at the vertices.

The method that is outlined here, in a restricted context, is obviously quite general and is applicable to anisotropic conductivity with heat sources as well as to the non-linear and convective systems discussed in Sections 4 and 5.

Boundary dissipation

When the boundary temperature of the solid is prescribed the variation $\delta\theta$ at the boundary vanishes. The value (2.33) of P_k is then reduced to

$$P_k = \iiint_{\tau} w \frac{\partial\theta}{\partial q_k}\, d\tau \tag{2.35 a}$$

and depends only on the thermal sources. This will also be the case if the temperature is prescribed on portions of the boundary while the remaining portions are impervious to heat. A formally similar case is obtained when the boundary heat transfer is represented by a surface heat transfer coefficient K. The boundary condition is then expressed by equation (2.1) of Chapter 2, which may be written

$$\mathbf{J}.\mathbf{n} = K(\theta-\theta_a), \tag{2.35 b}$$

where θ_a is the adiabatic surface temperature. If we introduce this value of $\mathbf{J}.\mathbf{n}$ into expression (2.33) for P_k, the latter becomes

$$P'_k = -\frac{1}{2}\frac{\partial}{\partial q_k}\iint_{A} K(\theta-\theta_a)^2\, dA + \iiint_{\tau} w\frac{\partial\theta}{\partial q_k}\, d\tau. \tag{2.35 c}$$

This leads to Lagrangian equations similar to equations (2.22). They are

$$\frac{\partial G}{\partial \dot{q}_k} + \frac{\partial D}{\partial q_k} = P_k. \tag{2.35 d}$$

However, the value of P_k is now given by expression (2.35 a), while D is replaced by

$$D = \frac{1}{2}\iiint_{\tau} \sum^{ij} k_{ij}\frac{\partial\theta}{\partial x_i}\frac{\partial\theta}{\partial x_j}\,d\tau + \frac{1}{2}\iint_{A} K(\theta-\theta_a)^2\,dA. \tag{2.35 e}$$

This value of the dissipation function is physically identical with expression (2.12) of Chapter 2 which includes the boundary dissipation.

3. OPERATIONAL PRINCIPLES

As already discussed in Chapter 3, a class of problems involving time operators may readily be formulated in an operational form by the use of simple operational rules. One such operational rule is

$$pf(t) = \frac{df(t)}{dt}, \tag{3.1}$$

where p is an operator that represents symbolically the derivative d/dt. We assume that the function $f(t)$ is zero for negative values of t and may be discontinuous at $t = 0$. The significance of the symbolic equation (3.1) results from the fact that it remains valid if we replace $f(t)$ and df/dt by their Laplace transforms. The symbol p then represents the variable of the Laplace transform. As already pointed out in Chapter 3, the validity of this rule requires the use of generalized functions corresponding to the discontinuities of the function $f(t)$. By this symbolism, equations involving time operators may be readily interpreted in terms of Laplace transforms.

By following the same procedures as in Section 6 of Chapter 3, the operational symbolism leads to operator-variational principles in complementary form. It must be assumed here that the thermal system is not only linear but also that its properties are independent of the time. In this context we shall consider the most general case with anisotropic thermal conductivity $k_{ij}(x, y, z)$ and heat capacity $c(x, y, z)$ both dependent on the coordinates. In addition, we assume distributed heat sources with a rate of heat generation $w(x, y, z, t)$ per unit volume dependent on the coordinates and the time.

In symbolic form, the law of heat conduction and the condition of energy conservation are written as

$$\begin{aligned} J_i &= -\sum^{j} k_{ij}\frac{\partial\theta}{\partial x_j}, \\ pc\theta &= -\sum^{i}\frac{\partial J_i}{\partial x_i} + w. \end{aligned} \tag{3.2}$$

In deriving the variational principle we proceed as if p were an algebraic quantity. We multiply the second of equations (3.2) by $\delta\theta$ and integrate over the volume τ. After integration by parts, we obtain

$$\iiint_\tau \left(pc\theta\,\delta\theta - \sum^i J_i \frac{\partial}{\partial x_i}\delta\theta\right) d\tau = -\iint_A \sum^i J_i n_i\,\delta\theta\,dA + \iiint_\tau w\,\delta\theta\,d\tau. \tag{3.3}$$

We denote by n_i the unit outward normal to the boundary A of the volume τ. We now substitute in the volume integral the value

$$J_i = -\sum^j k_{ij}(\partial\theta/\partial x_j).$$

We thus obtain the variational principle,

$$p\,\delta V + \delta D = -\iint_A \sum^i J_i n_i\,\delta\theta\,dA + \iiint_\tau w\,\delta\theta\,d\tau, \tag{3.4}$$

where

$$V = \tfrac{1}{2}\iiint_\tau c\theta^2\,d\tau,$$
$$D = \tfrac{1}{2}\iiint_\tau \sum^{ij} k_{ij}\frac{\partial\theta}{\partial x_i}\frac{\partial\theta}{\partial x_j}\,d\tau. \tag{3.5}$$

Equation (3.4) is the complementary form of the operator-variational principle (6.6) of Chapter 3. Its physical significance is the same as for equation (2.10). It provides an approximate verification of the condition of conservation of energy inside the volume τ.

The mathematical significance of the symbolic relation (3.4) is derived from the fact that the variables θ, J_i, and w may be replaced by their Laplace transforms. For example, θ may be replaced by its Laplace transform

$$\mathscr{L}(\theta) = \int_0^t e^{-pt'}\theta(t')\,dt'. \tag{3.6}$$

Actually, we need not introduce explicitly the symbol $\mathscr{L}$, so that the variables may be considered as representing either functions of time or Laplace transforms. As already pointed out in Section 6 of Chapter 3, this provides a very general and flexible symbolism which leads to a three-fold interpretation of operator-variational principles such as expressed by equation (3.4).

An operational interpretation is obtained by representing θ as‡

$$\theta = \sum^i q_i\,\theta_i(x,y,z). \tag{3.7}$$

‡ If the boundary temperature is prescribed we refer to the footnote remark in connection with equation (2.24) on p. 148. The expressions for V and D then become second degree polynomials in q_i.

From the values (3.5) and (3.7) we derive

$$V = \tfrac{1}{2}\sum^{ij} a_{ij}\,q_i\,q_j,$$
$$D = \tfrac{1}{2}\sum^{ij} b_{ij}\,q_i\,q_j. \tag{3.8}$$

The variational principle (3.4) leads to the equations

$$\frac{\partial}{\partial q_i}(pV+D) = P_i, \tag{3.9}$$

with
$$P_i = -\iint_A \sum^{k} J_k . n_k\,\theta_i\,dA + \iiint_\tau w\theta_i\,d\tau. \tag{3.10}$$

Substitution of the values (3.8) yields

$$\sum^{j} (pa_{ij}+b_{ij})q_j = P_i. \tag{3.11}$$

If we consider the variables to be functions of time, putting $p = d/dt$, we obtain equations (2.28) already derived above in the more restricted context of isotropic conductivity and in the absence of heat sources.

A second interpretation is obtained by considering the variational principle (3.4) to apply to the Laplace transforms themselves. In this case variations are applied to $\theta(x,y,z,p)$, where p is an algebraic parameter. The quantities J_i and w are also functions of p. The variational principle may then be applied for a sequence of fixed values of p.

A third interpretation is derived in terms of convolutions. This is based on the fact that symbolically the product of two Laplace transforms corresponds to a convolution. For example, in the expression

$$pV = \tfrac{1}{2}\iiint_\tau cp\theta^2\,d\tau \tag{3.12}$$

we may write
$$(p\theta)\theta = \int_0^t \theta(t-t')\dot{\theta}(t')\,dt'. \tag{3.13}$$

Similarly, other products of variables in equations (3.4) may be replaced by convolutions, leading to an interpretation of the variational principle in terms of functions of time.

Generalized operational method

The operational formulation of variational principles presented here and in Chapter 3 is obviously a particular case of a general procedure. It is generally applicable to problems governed by differential equations of the type

$$\sum^{i} \mathscr{P}_i(p)\mathscr{M}_i(x_k)f = 0, \tag{3.13 a}$$

where f is a function of a variable t and n variables x_k. We assume $\mathscr{M}_i(x_k)$ to be linear self-adjoint operators on the variables x_k, while $\mathscr{P}_i(p)$ are in symbolic form operators on the variable t and independent of t. By treating p as an algebraic quantity we readily derive variational principles as a consequence of the self-adjoint properties of $\mathscr{M}_i(x_k)$. The procedure is also valid for the case where equation (3.13a) is interpreted as a matrix equation.

4. CONDUCTION IN NON-LINEAR SYSTEMS

Consider a system where the parameters depend on the temperature θ. We may write, as before, the law of heat conduction

$$J_i = -\sum^{j} k_{ij}\frac{\partial\theta}{\partial x_j} \tag{4.1}$$

and the condition of energy conservation

$$c\dot{\theta} = -\sum^{i}\frac{\partial J_i}{\partial x_i}+w. \tag{4.2}$$

The heat capacity may now be a function of the coordinates and the temperature

$$c = c(x,y,z,\theta). \tag{4.3}$$

The thermal conductivity may be anisotropic and function of the coordinates, the time, and the temperature

$$k_{ij} = k_{ji} = k_{ij}(x,y,z,t,\theta). \tag{4.4}$$

If there are heat sources, the rate of heat generation per unit volume is $w(x,y,z,t)$. We multiply equation (4.2) by $\delta\theta$, and integrate the result over the volume τ. After integration by parts, we obtain

$$\iiint\limits_{\tau}\left(c\dot{\theta}\,\delta\theta-\sum^{i} J_i\frac{\partial\delta\theta}{\partial x_i}\right)d\tau = -\iint\limits_{A}\sum^{i} J_i n_i\,\delta\theta\,dA+\iiint\limits_{\tau} w\,\delta\theta\,d\tau. \tag{4.5}$$

In the volume integral we replace J_i by its value (4.1) derived from the law of thermal conduction. Hence

$$\iiint\limits_{\tau}\left(c\dot{\theta}\,\delta\theta+\sum^{ij} k_{ij}\frac{\partial\theta}{\partial x_j}\frac{\partial\delta\theta}{\partial x_i}\right)d\tau$$

$$= -\iint\limits_{A}\sum^{i} J_i n_i\,\delta\theta\,dA+\iiint\limits_{\tau} w\,\delta\theta\,d\tau. \tag{4.6}$$

As before, we introduce n generalized coordinates q_i and express θ as

$$\theta = \theta(q_1,q_2,\ldots,q_n,x,y,z,t). \tag{4.7}$$

The variations $\delta\theta$ are now expressed by means of the variations δq_k. Equations (2.16) and (2.19) remain valid for this case. Hence the variational equation (4.6) may be written as

$$\sum^{k}\left(\frac{\partial G}{\partial \dot{q}_k}+K_k\right)\delta q_k = P_k\,\delta q_k, \tag{4.8}$$

where $G = \frac{1}{2}\iiint_\tau c\dot{\theta}^2\,d\tau,$

$$K_k = \iiint_\tau \sum^{ij} k_{ij}\frac{\partial\theta}{\partial x_j}\frac{\partial^2\theta}{\partial x_i\,\partial q_k}, \tag{4.9}$$

$$P_k = -\iint_A \sum^{i} J_i n_i \frac{\partial\theta}{\partial q_k}\,dA + \iiint_\tau w\frac{\partial\theta}{\partial q_k}\,d\tau.$$

Since δq_k is arbitrary, we derive the n differential equations for q_k,

$$\frac{\partial G}{\partial \dot{q}_k}+K_k = P_k. \tag{4.10}$$

In expressions (4.9) c and k_{ij} are functions dependent on θ, and G is a second degree polynomial in $\dot{q}_k$ with coefficients dependent on q_k.

Temperature-independent conductivity

We consider the particular case where the thermal conductivity $k_{ij}(x, y, z, t)$ may depend on the coordinates and the time, but is independent of the temperature. If the heat capacity $c(x, y, z, \theta)$ depends on the temperature, the system remains non-linear. However, in this case we may write

$$K_k = \frac{\partial D}{\partial q_k} \tag{4.11}$$

with a dissipation function

$$D = \frac{1}{2}\iiint_\tau \sum^{ij} k_{ij}\frac{\partial\theta}{\partial x_i}\frac{\partial\theta}{\partial x_j}\,d\tau. \tag{4.12}$$

The differential equations (4.10) for the generalized coordinates become

$$\frac{\partial G}{\partial \dot{q}_k}+\frac{\partial D}{\partial q_k} = P_k. \tag{4.13}$$

They are identical in form to equations (2.22) for the linear case.

Case reducible to constant conductivity

This is the case where the thermal conductivity is of the form

$$k_{ij}(\theta) = k'_{ij}f(\theta), \tag{4.14}$$

with constant coefficients k'_{ij}. The heat capacity remains a function of the coordinates and the temperature. It was pointed out in Section 4 of Chapter 5 that this case is reducible to an analogue model with constant thermal conductivity, provided we introduce a fictitious temperature scale

$$u = \int_0^\theta f(\theta)\, d\theta. \tag{4.15}$$

With this variable u, equations (4.1) and (4.2) become

$$J_i = -\sum^j k'_{ij} \frac{\partial u}{\partial x_j},$$

$$c'\dot{u} = -\sum^i \frac{\partial J_i}{\partial x_i} + w, \tag{4.16}$$

where
$$c'(x, y, z, u) = \frac{1}{f(\theta)} c(x, y, z, \theta). \tag{4.17}$$

These equations govern thermal diffusion in a system with a temperature field u, a heat capacity c', and a constant thermal conductivity tensor k'_{ij}. Hence, in this analogue system, the generalized coordinates satisfy equations of the form (4.13) where

$$G = \tfrac{1}{2} \iiint_\tau c'\dot{u}^2\, d\tau,$$

$$D = \tfrac{1}{2} \iiint_\tau \sum^{ij} k'_{ij} \frac{\partial u}{\partial x_i} \frac{\partial u}{\partial x_j}\, d\tau, \tag{4.18}$$

$$P_k = -\iint_A \sum^i J_i n_i \frac{\partial u}{\partial q_k}\, dA + \iiint_\tau w \frac{\partial u}{\partial q_k}\, d\tau.$$

The heat flux J_i and the heat sources w retain their actual values.

5. CONVECTIVE SYSTEMS

Variational principles for heat convection in an incompressible fluid were discussed in Section 5 of Chapter 6 in the fundamental form based on heat displacement. Complementary principles may also be derived for this case.

The initial coordinates of the fluid particles are denoted by X_k. At the time t the coordinates of the fluid particles are

$$x_k = x_k(X_k, t). \tag{5.1}$$

The velocity field is

$$v_k = \frac{\partial x_k}{\partial t}. \tag{5.2}$$

This may be expressed as a function of the time and the coordinates x_k as

$$v_k = v_k(x_k, t). \tag{5.3}$$

With this expression for v_k, the condition of incompressibility of the fluid is

$$\sum^{k} \frac{\partial v_k}{\partial x_k} = 0. \tag{5.4}$$

The fluid thermal conductivity is

$$k_{ij} = k_{ji} = k_{ij}(x_k, t, \theta). \tag{5.5}$$

It may be a function of the coordinates, the time, and the temperature. The heat flux J_i satisfies the law of heat conduction

$$J_i = -\sum^{j} k_{ij} \frac{\partial \theta}{\partial x_j}. \tag{5.6}$$

The heat capacity of a fluid particle per unit volume is

$$c = c(X_k \theta), \tag{5.7}$$

and the heat content of the same particle per unit volume is

$$h = \int_0^{\theta} c\, d\theta = h(X_k, \theta). \tag{5.8}$$

The fluid may be heterogeneous and the fluid particles are tagged by their initial coordinates, X_k.

If we consider $\theta(x_k, t)$ to be an unknown function of x_k and t, the heat content h may also be considered a function of x_k and t as independent variables. With this understanding, conservation of energy is expressed by the equation

$$\frac{Dh}{Dt} + \sum^{i} \frac{\partial J_i}{\partial x_i} = w, \tag{5.9}$$

where

$$\frac{D}{Dt} = \frac{\partial}{\partial t} + \sum^{i} v_i \frac{\partial}{\partial x_i} \tag{5.10}$$

is the material derivative, and w represents the rate of heat generation per unit volume. Because of the condition of incompressibility (5.4), equation (5.9) coincides with equation (5.8) of Chapter 6. From the definition (5.8) of h we also derive

$$\frac{Dh}{Dt} = c\frac{D\theta}{Dt} = c\frac{\partial \theta}{\partial t} + c\sum^{i} v_i \frac{\partial \theta}{\partial x}. \tag{5.11}$$

Note that in the values (5.7) of c we may express X_k as a function of x_k and t. Hence, in equation (5.11) we may assume

$$c = c(x_k, t, \theta). \tag{5.12}$$

However, this is not an arbitrary function of the variables, since for θ constant it must satisfy the condition

$$\frac{Dc}{Dt} = \frac{\partial c}{\partial t} + \sum^{i} v_i \frac{\partial c}{\partial x_i} = 0. \tag{5.13}$$

A variational principle is obtained by multiplying by $\delta\theta$ the condition (5.9) for conservation of energy, and integrating the result over the volume τ. We obtain

$$\iiint_\tau \left(\frac{Dh}{Dt} + \sum^{i} \frac{\partial J_i}{\partial x_i}\right) \delta\theta \, d\tau = \iiint_\tau w \, \delta\theta \, d\tau. \tag{5.14}$$

Integration by parts, and the substitution of the value (5.6) for J_i in the volume integral, yields

$$\iiint_\tau \left(\frac{Dh}{Dt}\delta\theta + \sum^{ij} k_{ij} \frac{\partial\theta}{\partial x_j} \frac{\partial\delta\theta}{\partial x_i}\right) d\tau$$

$$= -\iint_A \sum^{i} J_i n_i \, \delta\theta \, dA + \iiint_\tau w \, \delta\theta \, d\tau. \tag{5.15}$$

With arbitrary values of $\delta\theta$ this equation provides a variational procedure for verifying conservation of energy.

As before, we may consider θ to be defined in terms of n generalized coordinates q_i. We write

$$\theta = \theta(q_1, q_2, \ldots, q_n, x_k, t). \tag{5.16}$$

The variations $\delta\theta$ are then expressed in terms of δq_i by equations (2.16) and (2.19). We substitute these values of $\delta\theta$ into equation (5.15) and introduce expression (5.11) for Dh/Dt. The variational equation (5.15) becomes

$$\sum^{k} \left(\frac{\partial G}{\partial \dot{q}_k} + L_k + K_k\right) \delta q_k = \sum^{k} P_k \, \delta q_k, \tag{5.17}$$

where G, K_k, and P_k are the same as in equations (4.9), and

$$L_k = \iiint_\tau c \sum^{i} v_i \frac{\partial\theta}{\partial x_i} \frac{\partial\theta}{\partial q_k} \, d\tau. \tag{5.18}$$

Since δq_k are arbitrary, the generalized coordinates obey the n differential equations,

$$\frac{\partial G}{\partial \dot{q}_k} + L_k + K_k = P_k. \tag{5.19}$$

In the particular case where the thermal conductivity $k_{ij}(x,y,z,t)$ is independent of the temperature, the value of K_k is given by expression (4.11) and the differential equations (5.9) become

$$\frac{\partial G}{\partial \dot{q}_k}+\frac{\partial D}{\partial q_k}+L_k = P_k. \tag{5.20}$$

The equations are also reducible to this form if the thermal conductivity remains a function of the temperature given by the expression

$$k_{ij}(\theta) = k'_{ij}\,f(\theta), \tag{5.21}$$

where k'_{ij} are constants. As shown in the preceding section, this is accomplished by a change of the temperature scale.

Mixed solid–fluid systems and turbulent flow

The case of pure conduction in a fixed solid in a particular case of equations (5.19) obtained by putting $v_i = 0$. Hence, in this case $L_k = 0$, and the result coincides with equation (4.10). Equations (5.19) are therefore applicable to a mixed system composed of solids and moving fluids. In the solid regions we put $v_i = 0$. In regions of turbulent flow k_{ij} is replaced by the total conductivity

$$K_{ij} = k\,\delta_{ij}+c\epsilon_{ij}, \tag{5.22}$$

where δ_{ij} is the Kronecker symbol, k is the molecular conductivity, and ϵ_{ij} the turbulent diffusivity.

Operational form

For a linear convective system with time-independent parameters, complementary principles may also be expressed in operational form. In this case we must assume c, k_{ij}, v_i to depend only on the coordinates. Hence the velocity field is independent of time. We replace $\partial/\partial t$ by the operator p in equation (5.9). Equations (5.6) and (5.9) become

$$J_i = -\sum^{j} k_{ij}\frac{\partial \theta}{\partial x_j}, \qquad pc\theta+c\sum^{i} v_i\frac{\partial \theta}{\partial x_i}+\sum^{i}\frac{\partial J_i}{\partial x_i} = w. \tag{5.23}$$

We then proceed as in Section 3 for the case of pure conduction and derive

$$p\,\delta V+\delta D+\iiint_\tau c\sum^{i} v_i\frac{\partial \theta}{\partial x_i}\,\delta\theta\,d\tau = -\iint_A \sum^{i} J_i n_i\,\delta\theta\,dA+\iiint_\tau w\,\delta\theta\,d\tau, \tag{5.24}$$

where V and D are defined by equations (3.5). As explained in Section 3, the operational form (5.24) of the variational principle may be interpreted in three different ways on the basis of Laplace transforms. They include an algebraic interpretation and another in terms of convolutions. When θ is expressed as

$$\theta = \sum^{i} q_i \theta_i(x, y, z), \tag{5.25}$$

the variational principle (5.24) yields the equations

$$\frac{\partial}{\partial q_k}(pV+D)+L_k = P_k, \tag{5.26}$$

where

$$\begin{aligned} L_k &= \iiint_\tau c \sum^{i} v_i \frac{\partial \theta}{\partial x_i} \theta_k \, d\tau, \\ P_k &= -\iint_A \sum^{i} J_i n_i \theta_k \, dA + \iiint_\tau w\theta_k \, d\tau. \end{aligned} \tag{5.27}$$

The operational interpretation of equations (5.26), putting $p = d/dt$, yields a set of linear differential equations for q_k.

APPENDIX

RELATED SUBJECTS

1. INTRODUCTION

In the eight chapters that constitute the main body of this book the material is presented exclusively in the context of heat transfer. However, the methods and concepts involved are of much wider scope. They are applicable to a large category of phenomena that involve energy dissipation and include mass and energy transport, inertial dynamics, electrodynamics, and irreversible thermodynamics. In addition, this book also constitutes an attempt to restore the Lagrangian viewpoint in physics by showing that it is part of a general framework and a powerful tool of analysis, which leads to a unified outlook and brings out generalizations, hidden common properties, and analogies between widely different types of phenomena. The purpose of this appendix is to cast the material in this more general perspective.

Mass transport problems are discussed in Section 2: in particular, reference is made to isothermal diffusion and convection of solutions in physical chemistry, and problems of fluid- and moisture-seepage through porous media. Possible applications to neutron diffusion and to the dynamics of nuclear reactors is also indicated.

A broader physical outlook is provided through the Lagrangian analysis of irreversible thermodynamics. Actually, the variational treatment of heat transfer was suggested to the author as an outgrowth of more general thermodynamic principles for irreversible processes and not through any mathematical formalism. The particular subject of thermoelasticity is presented in Section 3 as an illustration of Lagrangian thermodynamics. Brief mention is made of applications to viscoelasticity and the dynamics of viscous fluids.

An important step is also to clarify the significance of the Lagrangian approach from the purely mathematical viewpoint of functional analysis. This is the object of Section 4. The physical relevance of much of the formalism of functional analysis is lost in the context of the theory of sets. It has been our purpose in this book to restore some of the physical relevance by presenting the fundamental mathematical ideas in a physical context. It has become fashionable to translate the concept of virtual work and derived forms into the language of functional space. In fact, such concepts are not new and have been used by physicists and engineers for many generations, particularly in the field of classical mechanics. An abstract and very general concept that embodies the essential features may be called the *variational scalar product*. It brings into a unified framework various methods such as the more traditional form of the variational calculus, as well as others that are usually disguised in the literature under different names.

It is also pointed out that the continuous mathematical model of a physical system is quite artificial and introduces many spurious difficulties. The concept of *resolution threshold* justifies the use of a finite number of generalized coordinates as a description that is accurate and complete from the physical standpoint. There is no point in reaching beyond this threshold, because the continuous mathematical model becomes inadequate in the microscopic scale.

2. MASS TRANSPORT

Most problems of mass transport obey the same equations as those that govern heat transfer. The variational principles and Lagrangian equations developed in this book in the context of heat transfer are therefore immediately applicable to their mathematical analogue for mass transport by a simple change of notation. As an illustration, we shall briefly discuss three types of mass transport phenomena to which the methods are applicable.

Isothermal diffusion

We consider the diffusion of a molecular or atomic species through a medium at rest and at uniform temperature. The rate of diffusion depends on the concentration gradient and is expressed by Fick's law,

$$\dot{\mathbf{M}} = -\mathscr{D}\,\mathrm{grad}\,m. \tag{2.1}$$

In this equation m is the concentration defined as the mass of the diffusing species per unit volume. The vector $\mathbf{M}$ is the mass displacement and its time derivative $\dot{\mathbf{M}}$ is the rate of diffusion. The coefficient of diffusion denoted by $\mathscr{D}$ may be a function of the concentration m. Conservation of mass is expressed by the relation

$$m = -\mathrm{div}\,\mathbf{M}. \tag{2.2}$$

Combining equations (2.1) and (2.2) we derive the diffusion equation

$$\frac{\partial m}{\partial t} = \operatorname{div}(\mathscr{D} \operatorname{grad} m). \tag{2.3}$$

These equations are identical to equations (2.2), (2.4), and (2.10 a) of Chapter 1 for the linear case, provided we put

$$\begin{aligned} c &= 1, \quad \mathbf{M} = \mathbf{H}, \\ m &= \theta, \quad \mathscr{D} = k. \end{aligned} \tag{2.4}$$

For the non-linear case when $\mathscr{D}$ is a function of m the isothermal diffusion problem is governed by the same equations as in the non-linear heat conduction discussed in Chapter 5. In addition to relations (2.4), we also put

$$m = \theta = h \tag{2.5}$$

in equations (2.1) and (2.2) of that chapter.

The variational principles derived in this book are therefore valid for isothermal diffusion. In particular, the mass displacement $\mathbf{M}$, which is the analogue of the heat displacement $\mathbf{H}$, may be expressed in terms of generalized coordinates q_i as

$$\mathbf{M} = \mathbf{M}(q_1, q_2, \ldots, q_n, x, y, z, t). \tag{2.6}$$

By varying this field, subject to the constraint equation (2.2), we derive the Lagrangian equations,

$$\frac{\partial V}{\partial q_i} + \frac{\partial D}{\partial \dot{q}_i} = Q_i. \tag{2.7}$$

In these equations the thermal potential is replaced by

$$V = \tfrac{1}{2} \iiint\limits_{\tau} m^2 \, d\tau, \tag{2.8}$$

where τ is the volume of the medium. The dissipation function is

$$D = \frac{1}{2} \iiint\limits_{\tau} \frac{1}{\mathscr{D}} \dot{\mathbf{M}}^2 \, d\tau. \tag{2.9}$$

The generalized force is

$$Q_i = -\iint\limits_{A} m \frac{\partial \mathbf{M}}{\partial q_i} \cdot \mathbf{n} \, dA, \tag{2.10}$$

where $\mathbf{n}$ is the unit outward normal at the boundary A of the volume τ.

The case of simultaneous mass convection and diffusion in a moving incompressible fluid at uniform temperature is entirely similar to that of convective heat transfer treated in Chapter 6. Fick's law (2.1) becomes

$$\mathbf{J} = -\mathscr{D} \operatorname{grad} m. \tag{2.11}$$

The rate of diffusion is now

$$\mathbf{J} = \dot{\mathbf{M}} - \mathbf{v}m, \tag{2.12}$$

where $\mathbf{v}$ is the velocity field of the fluid. The mass displacement $\mathbf{M}$ is the sum of the convective and diffusive transport. The mass conservation equation (2.2) remains valid. Combining equations (2.2), (2.11), and (2.12) and taking into account the condition of incompressibility

$$\operatorname{div} \mathbf{v} = 0, \tag{2.13}$$

yields for the mass concentration m the field equation

$$\frac{\partial m}{\partial t} = \operatorname{div}(\mathscr{D} \operatorname{grad} m) - \mathbf{v} . \operatorname{grad} m. \tag{2.14}$$

The Lagrangian equations (2.7) are applicable to the case of convective mass transport, provided the dissipation function is defined as

$$D = \frac{1}{2} \iiint_{\tau} \frac{1}{\mathscr{D}} \mathbf{J}^2 \, d\tau. \tag{2.15}$$

This corresponds to the dissipation function (5.23) of Chapter 6 for the case of heat transfer. These results also include the case of mass transport in turbulent flow. In this case we must replace $\mathscr{D}$, in the equations above, by

$$\mathscr{A} = \mathscr{D} + \epsilon, \tag{2.16}$$

where ϵ is the turbulent diffusivity.

For anisotropic diffusion the treatment is identical to the case of anisotropic thermal conductivity and turbulence, considered in Section 5 of Chapter 6.

Seepage through porous media

In a wide category of technological problems concerned with ground water flow, seepage through earth dams, and oil exploitation, we have to solve the equations of motion of a fluid in a partially saturated porous medium. Such equations are similar to the diffusion equation (2.3). In the case of water seepage, m plays the role of the moisture content. The diffusion coefficient $\mathscr{D}$ is strongly dependent on the moisture content and the problems are therefore essentially non-linear. In the general case, a term must be added to equation (2.3) in order to account for the action of both capillary and gravity forces. The variational principles and Lagrangian equations developed in this book are readily applicable to such non-linear seepage problems.

Neutron diffusion and nuclear reactors

Similar procedures may also be used for the approximate analysis of nuclear reactor dynamics. While the application remains straightforward, appropriate modifications must be introduced because the physics involves not only diffusion but also neutron generation. This corresponds to the presence of sources depending on the neutron concentration. In addition, account must be taken of delayed neutron emission which introduces a relaxation time in the sources. This effect also may be incorporated into the variational and Lagrangian treatment.

3. IRREVERSIBLE THERMODYNAMICS

A very general Lagrangian treatment of irreversible processes was originally developed by the author in 1954.‡ A more extensive discussion on the implications of this theory was provided in a later paper.§ This formulation of irreversible thermodynamics by Lagrangian equations, and its corresponding variational principles, are based on the introduction of a new thermodynamic potential for systems that are not at uniform temperature and of a dissipation function derived from Onsager's relations. Domains of applications embrace a wide range of phenomena in the mechanics of viscous and viscoelastic media, as well as in thermophysics, physical chemistry, and electrodynamics.

Actually, this development preceded the application to heat transfer as presented in this book. It is the basic thermodynamics and, in particular, its application to thermoelasticity, which provided the key for the discovery of variational principles in heat transfer. The Lagrangian theory of thermoelasticity based on irreversible thermodynamics was introduced by the author in 1956.|| The resulting reciprocity relations and their applications to structural analysis were also discussed by the author‡‡ and by Kowalewski.§§ A detailed discussion of

‡ M. A. Biot, 'Theory of stress–strain relations in anisotropic viscoelasticity and relaxation phenomena', *J. Appl. Phys.* **25**, 1385–91 (1954).

§ M. A. Biot, 'Linear thermodynamics and the mechanics of solids', *Proceedings of the Third U.S. National Congress of Applied Mechanics*, pp. 1–18. American Society of Mechanical Engineers, New York (1958).

|| M. A. Biot, 'Thermoelasticity and irreversible thermodynamics', *J. appl. Phys.* **27**, 240–53 (1956).

‡‡ M. A. Biot, 'New thermomechanical reciprocity relations with applications to thermal stress analysis', *J. Aerospace Sci.* **26**, 401–8 (1959).

§§ J. Kowalewski, 'Influence functions for displacements and stresses from temperature and heat loads', *AIAA Jl*, **5**, 1694–6 (1967).

thermoelasticity from the Lagrangian viewpoint is also found in a recent paper by Rafalski.‡

In the following, as an illustration of the more general thermodynamic viewpoint, we present an outline restricted to the particular case of thermoelasticity, along with a short reference to Lagrangian methods in the dynamics of viscous fluids, viscoelasticity, and porous media, and in electrodynamics.

Thermoelasticity

We consider an anisotropic elastic medium initially in mechanical and thermodynamic equilibrium at a uniform absolute 'reference temperature' T_r. A small disturbance from this equilibrium state is defined by a displacement field u_i of the solid and a temperature field θ representing the deviation from the initial value T_r. With Cartesian coordinates x_i, the temperature, the strain $e_{ij} = \frac{1}{2}(\partial u_i/\partial x_j + \partial u_j/\partial x_i)$, and the stress σ_{ij} are related by the following equations:

$$\sigma_{\mu\nu} = \sum^{ij} C^{ij}_{\mu\nu} e_{ij} - \beta_{\mu\nu}\theta, \tag{3.1}$$

where $\beta_{\mu\nu}$ and $C^{ij}_{\mu\nu}$ are physical constants. The latter are the twenty-one isothermal elastic moduli. These constants satisfy the following symmetry relations:

$$\begin{aligned} \beta_{\mu\nu} &= \beta_{\nu\mu}, \\ C^{ij}_{\mu\nu} &= C^{\mu\nu}_{ij} = C^{ji}_{\mu\nu} = C^{ij}_{\nu\mu}. \end{aligned} \tag{3.2}$$

From classical thermodynamics we also derive the following relation,

$$s = \frac{c\theta}{T_r} + \sum^{ij} \beta_{ij} e_{ij}, \tag{3.3}$$

where s is the increment of entropy per unit volume and c the specific heat per unit volume under zero strain. We may solve equation (3.3) for θ and substitute this value into equations (3.1). We obtain

$$\begin{aligned} \sigma_{\mu\nu} &= \sum^{ij} \left(C^{ij}_{\mu\nu} + \frac{T_r}{c}\beta_{\mu\nu}\beta_{ij}\right) e_{ij} - \frac{T_r}{c}\beta_{\mu\nu} s, \\ \theta &= -\frac{T_r}{c}\sum^{ij} \beta_{ij} e_{ij} + \frac{T_r}{c} s. \end{aligned} \tag{3.4}$$

These relations lead to the following important result.

Consider the quadratic invariant

$$v = \tfrac{1}{2}\sum^{ij} \sigma_{ij} e_{ij} + \tfrac{1}{2}\theta s. \tag{3.5}$$

‡ P. Rafalski, 'The Lagrangian formulation of the dynamic thermoelastic problem for mixed boundary conditions', *Proc. Vibr. Probl.* **9**, 17–35 (1968).

With the values (3.4) for σ_{ij} and θ, it is a function of the independent variable e_{ij} and s. Because the matrix of coefficients in equations (3.4) is symmetric, we may write the following property:

$$\sigma_{ij} = \frac{\partial v}{\partial e_{ij}}, \qquad \theta = \frac{\partial v}{\partial s}. \tag{3.6}$$

In the partial derivations the variables e_{ij} are treated as nine independent variables. The volume integral

$$V = \iiint_{\tau} v \, d\tau \tag{3.7}$$

was introduced by the author as the *thermoelastic potential* of the domain τ. The invariant v is therefore the specific thermoelastic potential per unit volume.

We may also express v in terms of e_{ij} and θ as independent variables. We obtain

$$v = \tfrac{1}{2} \sum^{ij} \sum^{\mu\nu} C^{ij}_{\mu\nu} e_{ij} e_{\mu\nu} + \frac{1}{2} \frac{c}{T_r} \theta^2. \tag{3.8}$$

For the domain τ we write

$$V = V_r + \frac{1}{2T_r} \iiint c\theta^2 \, d\tau,$$

with

$$V_r = \tfrac{1}{2} \iiint_{\tau} \sum^{ij} \sum^{\mu\nu} C^{ij}_{\mu\nu} e_{ij} e_{\mu\nu}. \tag{3.9}$$

For isothermal deformations ($\theta = 0$) the potential V reduces to V_r which is the classical free energy of the system. It coincides with the familiar strain energy of the theory of elasticity.

As shown by the author, the thermoelastic potential V is a particular case of a more general thermodynamic potential for systems with a distributed non-uniform temperature. It coincides with the free-energy if the temperature is uniform.

For our purpose we must introduce an *entropy displacement* vector defined as

$$S_i = \frac{1}{T_r} H_i. \tag{3.10}$$

The heat displacement is denoted by H_i. In terms of S_i the entropy density s becomes

$$s = -\sum^{i} \frac{\partial S_i}{\partial x_i}. \tag{3.11}$$

Hence, equation (3.3) may be written

$$\theta = -\frac{T_r}{c} \sum^{i} \frac{\partial S_i}{\partial x_i} - \frac{T_r}{c} \sum^{ij} \beta_{ij} e_{ij}. \tag{3.12}$$

For a system that is not in thermodynamic and mechanical equilibrium we must add the equation of motion,

$$\sum^{j} \frac{\partial \sigma_{ij}}{\partial x_j} = \rho \ddot{u}_i, \tag{3.13}$$

and the law of heat conduction,

$$\dot{H}_i = -\sum^{j} k_{ij} \frac{\partial \theta}{\partial x_j}. \tag{3.14}$$

In these equations ρ is the mass density and k_{ij} the thermal conductivity. For our purpose, equation (3.14) is written in the equivalent form

$$\frac{\partial \theta}{\partial x_i} = -T_r \sum^{j} \lambda_{ij} \dot{S}_j, \tag{3.15}$$

where the thermal resistivity λ_{ij} is defined by the inverse matrix of k_{ij}.

With these definitions it is possible to derive a variational principle. The unknown variables describing the physical system are chosen to be u_i and S_i. Hence expression (3.7) for V is also a function of these variables by substituting the values (3.4), (3.11), and (3.12) for σ_{ij}, s, and θ. The variational principle is

$$\delta V + \iiint_\tau \rho \sum^{i} \ddot{u}_i \, \delta u_i \, d\tau + T_r \iiint_\tau \sum^{ij} \lambda_{ij} \dot{S}_j \, \delta S_i \, d\tau$$
$$= \iint_A \Big(\sum^{i} F_i \, \delta u_i - \theta \sum^{i} n_i \, \delta S_i \Big) \, dA. \tag{3.16}$$

In this equation δu_i and δS_i are arbitrary variations, F_i is the force per unit area applied at the boundary A of the volume τ, and n_i is the unit outward normal to the boundary.

The variational principle (3.16) is derived as follows. The variation of the thermoelastic potential (3.7) is

$$\delta V = \iiint_\tau \delta v \, d\tau, \tag{3.17}$$

where $$\delta v = \tfrac{1}{2} \sum^{ij} \sigma_{ij} \, \delta e_{ij} + \tfrac{1}{2} \sum^{ij} e_{ij} \, \delta \sigma_{ij} + \tfrac{1}{2} \theta \, \delta s + \tfrac{1}{2} s \, \delta \theta. \tag{3.18}$$

Substitution of expressions (3.6) into the value (3.18) of δv yields

$$\delta v = \sum^{ij} e_{ij} \, \delta \sigma_{ij} + s \, \delta \theta. \tag{3.19}$$

Hence equation (3.18) may be also written as

$$\delta v = \sum^{ij} \sigma_{ij} \, \delta e_{ij} + \theta \, \delta s \tag{3.20}$$

and the variation δV becomes

$$\delta V = \iiint_{\tau} \left(\sum^{ij} \sigma_{ij}\,\delta e_{ij} + \theta\,\delta s\right) d\tau. \tag{3.21}$$

We note that the strain is defined as

$$e_{ij} = \frac{1}{2}\left(\frac{\partial u_i}{\partial x_j} + \frac{\partial u_j}{\partial x_i}\right), \tag{3.22}$$

and s is given by expression (3.11). The volume integral (3.21) is integrated by parts and introduced into the variational equation (3.16). We obtain

$$\iiint_{\tau} \sum^{ij} \left(\rho \ddot{u}_i - \frac{\partial \sigma_{ij}}{\partial x_j}\right) \delta u_i\, d\tau + \iiint_{\tau} \sum^{ij} \left(\frac{\partial \theta}{\partial x_i} + T_r \lambda_{ij} \dot{S}_j\right) \delta S_i\, d\tau = 0. \tag{3.23}$$

This relation is verified for all variations δu_i and δS_i if the equations of motion (3.13) and the equations of heat conduction (3.15) are verified. Hence the variational principle (3.16) is nothing but another form of these equations.

The derivation of Lagrangian equations from the variational principle (3.16) is carried out as follows. We express the unknown variables u_i and S_i in terms of n generalized coordinates $q_1, q_2, \ldots, q_n$. We write

$$u_i = u_i(q_1, q_2, \ldots, q_n, x, y, z, t), \qquad S_i = S_i(q_1, q_2, \ldots, q_n, x, y, z, t). \tag{3.24}$$

The generalized coordinates are now given arbitrary variations δq_i. Hence

$$\delta u_i = \sum^{k} \frac{\partial u_i}{\partial q_k}\,\delta q_k, \qquad \delta S_i = \sum^{k} \frac{\partial S_i}{\partial q_k}\,\delta q_k. \tag{3.25}$$

The surface integral in equation (3.16) may therefore be written as

$$\iint_{A} \left(\sum^{i} F_i\,\delta u_i - \theta \sum^{i} n_i\,\delta S_i\right) dA = \sum^{k} Q_k\,\delta q_k, \tag{3.26}$$

where

$$Q_k = \iint_{A} \left(\sum^{i} F_i \frac{\partial u_i}{\partial q_k} - \theta \sum^{i} n_i \frac{\partial S_i}{\partial q_k}\right) dA. \tag{3.27}$$

Also

$$\delta V = \sum^{k} \frac{\partial V}{\partial q_k}\,\delta q_k. \tag{3.28}$$

The term

$$\iiint_{\tau} \rho \sum^{i} \ddot{u}_i\,\delta u_i = \iiint_{\tau} \rho \sum^{ik} \ddot{u}_i \frac{\partial u_i}{\partial q_k}\,\delta q_k \tag{3.29}$$

may be expressed by means of the kinetic energy, using a classical procedure of Lagrangian mechanics. To show this, we start from the identity

$$\ddot{u}_i \frac{\partial u_i}{\partial q_k} = \frac{d}{dt}\left(\dot{u}_i \frac{\partial u_i}{\partial q_k}\right) - \dot{u}_i \frac{d}{dt}\left(\frac{\partial u_i}{\partial q_k}\right). \tag{3.30}$$

From equations (3.24) we derive the relation

$$\dot{u}_i = \sum^{k} \frac{\partial u_i}{\partial q_k} \dot{q}_k + \frac{\partial u_i}{\partial t}. \tag{3.31}$$

Hence

$$\frac{\partial \dot{u}_i}{\partial \dot{q}_k} = \frac{\partial u_i}{\partial q_k} \tag{3.32}$$

and

$$\frac{\partial \dot{u}_i}{\partial q_k} = \sum^{j} \frac{\partial^2 u_i}{\partial q_k \partial q_j} \dot{q}_j + \frac{\partial^2 u_i}{\partial q_k \partial t} = \frac{d}{dt}\left(\frac{\partial u_i}{\partial q_k}\right). \tag{3.33}$$

With these values, the identity (3.30) becomes

$$\ddot{u}_i \frac{\partial u_i}{\partial q_k} = \frac{d}{dt}\left(\dot{u}_i \frac{\partial \dot{u}_i}{\partial \dot{q}_k}\right) - \dot{u}_i \frac{\partial \dot{u}}{\partial q_k}. \tag{3.34}$$

From this result we derive

$$\iiint_{\tau} \rho \sum^{i} \ddot{u}_i \, \delta u_i = \sum^{k} \left\{\frac{d}{dt}\left(\frac{\partial \mathscr{T}}{\partial \dot{q}_k}\right) - \frac{\partial \mathscr{T}}{\partial q_k}\right\} \delta q_k, \tag{3.35}$$

where $\mathscr{T}$ is the kinetic energy

$$\mathscr{T} = \tfrac{1}{2} \iiint_{\tau} \rho \sum^{i} \dot{u}_i^2 \, d\tau. \tag{3.36}$$

By an entirely similar procedure we may also transform the remaining term in equation (3.16). From expressions (3.24) we derive

$$\dot{S}_i = \sum^{k} \frac{\partial S_i}{\partial q_k} \dot{q}_k + \frac{\partial S_i}{\partial t}. \tag{3.37}$$

Hence

$$\frac{\partial \dot{S}_i}{\partial \dot{q}_k} = \frac{\partial S_i}{\partial q_k}. \tag{3.38}$$

Therefore we may write

$$T_r \iiint_{\tau} \sum^{ij} \lambda_{ij} \dot{S}_j \, \delta S_i \, d\tau = \sum^{k} T_r \iiint_{\tau} \sum^{ij} \lambda_{ij} \dot{S}_j \frac{\partial \dot{S}_i}{\partial \dot{q}_k} \, d\tau \, \delta q_k \tag{3.39}$$

or

$$T_r \iiint_{\tau} \sum^{ij} \lambda_{ij} \dot{S}_j \, \delta S_i \, d\tau = \sum^{k} \frac{\partial D}{\partial \dot{q}_k} \delta q_k, \tag{3.40}$$

with

$$D = \tfrac{1}{2} T_r \iiint_{\tau} \sum^{ij} \lambda_{ij} \dot{S}_i \dot{S}_j \, d\tau. \tag{3.41}$$

In this derivation we have assumed that Onsager's reciprocity relations

$$\lambda_{ij} = \lambda_{ji} \tag{3.42}$$

are satisfied. The invariant D is a dissipation function. As shown in earlier work,‡ the thermodynamic significance of the dissipation function is provided by the fact that $2D/T_r$ is equal to the rate of entropy production per unit volume.

Substitution of the expressions (3.26), (3.28), (3.35), and (3.40) into the variational principle (3.16) yields

$$\sum^{k} \left\{\frac{d}{dt}\left(\frac{\partial \mathscr{T}}{\partial \dot{q}_k}\right) - \frac{\partial \mathscr{T}}{\partial q_k} + \frac{\partial D}{\partial \dot{q}_k} + \frac{\partial V}{\partial q_k}\right\}\delta q_k = \sum^{k} Q_k\, \delta q_k. \tag{3.43}$$

Since the δq_ks are arbitrary, we derive

$$\frac{d}{dt}\left(\frac{\partial \mathscr{T}}{\partial \dot{q}_k}\right) - \frac{\partial \mathscr{T}}{\partial q_k} + \frac{\partial D}{\partial \dot{q}_k} + \frac{\partial V}{\partial q_k} = Q_k. \tag{3.44}$$

These are the Lagrangian equations for the generalized coordinates q_k.

By proceeding as in Section 2 of Chapter 2, these equations may include a boundary condition characterized by a surface heat-transfer coefficient K. This is obtained by adding to the dissipation function a term representing the boundary dissipation. We write

$$D = \tfrac{1}{2}T_r \iiint\limits_{\tau} \sum^{ij} \lambda_{ij}\, \dot{S}_i\, \dot{S}_j\, d\tau + \tfrac{1}{2}T_r \iint\limits_{A} \frac{1}{K}\, \dot{S}_n^2\, dA, \tag{3.45}$$

where S_n is the normal component of the entropy displacement at the boundary A. The thermal force must also be replaced by

$$Q_k = \iint\limits_{A} \left(\sum^{i} F_i \frac{\partial u_i}{\partial q_k} - \theta_a \sum^{i} n_i \frac{\partial S_i}{\partial q_k}\right) dA. \tag{3.46}$$

The quantity θ_a is the adiabatic surface temperature as defined in Section 2 of Chapter 2.

Thermal conduction as a particular case

These general results are applicable to pure thermal conduction. This is obtained by putting $u_i = 0$. The value (3.5) of v may then be written as

$$v = \tfrac{1}{2}\theta s = \frac{1}{2}\frac{c\theta^2}{T_r}. \tag{3.47}$$

A similar expression is obtained if we put $\sigma_{ij} = 0$. In this case c represents

‡ See p. 165, nn. § and ||.

the heat capacity under zero stress. Hence, for pure conduction, the thermoelastic potential becomes

$$V = \frac{1}{2T_r}\iiint_\tau c\theta^2\,d\tau. \tag{3.48}$$

In the pure conduction problem the kinetic energy $\mathscr{T}$ is either zero or negligible. Hence, putting $\mathscr{T} = 0$, the Lagrangian equations (3.44) become

$$\frac{\partial V}{\partial q_k}+\frac{\partial D}{\partial \dot{q}_k} = Q_k. \tag{3.49}$$

Except for a factor $1/T_r$, the expressions V, D, and Q_k coincide respectively with the thermal potential, the dissipation function, and the thermal force defined in Chapters 1 and 2. The Lagrangian equations for pure conduction derived in that chapter are identical to equations (3.49) after cancellation of the common factor $1/T_r$.

Similarly, by neglecting the displacements and accelerations, the variational principle (3.16) is found to coincide with the variational principle (5.10) of Chapter 1.

Viscous fluids, viscoelasticity, and porous media

The Lagrangian equations (3.44) are applicable to the dynamics of an incompressible viscous fluid. The motion of the fluid is described by generalized coordinates and the viscosity is taken into account by the use of the well-known dissipation function of Rayleigh. Such a procedure yields a powerful tool for the analysis of many problems of fluid dynamics. Of particular interest are the possible applications to boundary layer flow. The dynamics of viscous fluids and viscoelastic media under initial stress based on Lagrangian equations has been treated in a book‡ by the author. The mechanics of porous media containing a viscous fluid has also been developed by the use of Lagrangian equations for elastic and viscoelastic solids.§ The case of an initially stressed porous medium has also been considered‖ by applying the same methods. The common underlying thermodynamic background of these various theories brings to light a fundamental analogy between thermoelasticity and the mechanics of porous media.

‡ M. A. Biot, *Mechanics of incremental deformations*', Chapter 6. Wiley, New York (1965).

§ M. A. Biot, 'Generalized theory of acoustic propagation in porous dissipative media', *J. acoust. Soc. Am.* **34,** 1254–64 (1962).

‖ M. A. Biot, 'Theory of stability and consolidation of a porous medium under initial stress', *J. Math. Mech.* **12,** 521–42 (1963).

Electrodynamics

The possibility of expressing the equations of electrodynamics in Lagrangian form has long been recognized. The required formalism involves a dissipation function that expresses the Joule effect. A unified Lagrangian approach of coupled electrical and mechanical systems is particularly useful in the field of electroacoustics.‡ In an attempt to derive general principles, a Lagrangian formulation of electromagnetic theory and fluid dynamics has also been discussed in two papers by Herivel.§

4. GENERALIZED COORDINATES AND FUNCTIONAL ANALYSIS

The concept of generalized coordinates and the corresponding Lagrangian equations were already an essential part of the early developments of classical mechanics. The important addition of the dissipation function by Rayleigh extended these concepts to systems including viscous forces. While generalized coordinates have been almost exclusively applied to problems of mechanics, the concept and the associated methods have a much wider scope and provide the foundation for the analysis of a large category of phenomena. From a purely mathematical standpoint, the Lagrangian methods also suggest new viewpoints when interpreted in the context of functional analysis. In order to clarify this broader outlook, let us examine more closely the fundamental aspects of the Lagrangian procedure.

The standard derivation in classical mechanics starts with d'Alembert's principle based on the concept of virtual work. When applied to a system of n mass points it is expressed in the form

$$\sum^{k} m_k \ddot{\mathbf{r}}_k . \delta\mathbf{r}_k - \sum^{k} \mathbf{F}_k . \delta\mathbf{r}_k = 0. \tag{4.1}$$

The displacement of the mass m_k is $\mathbf{r}_k$ and the force acting on it is $\mathbf{F}_k$. The summation is extended to the n masses. Equation (4.1) is valid for arbitrary variations $\delta\mathbf{r}_k$. The remarkable power and flexibility of this formulation of mechanics is based on several fundamental properties. For example, in the case of holonomic constraints certain types of unknown constraint forces vanish automatically from the equations

‡ For an application of Lagrangian equations to electroacoustics see, for example, Th. v. Kármán and M. A. Biot, *Mathematical methods in engineering*, chapter 6, p. 256. McGraw-Hill, New York (1940).

§ J. W. Herivel, 'A general variational principle for dissipative systems', *Proc. R. Ir. Acad.* **56**, Sect. A, 37–44, 67–75 (1954).

because their virtual work is zero. Furthermore, the variations may be chosen to satisfy the constraints. This leads to equations that contain only as many unknowns as there are degrees of freedom and from which constraint forces are eliminated. Furthermore, it is usually possible to express some or all of the terms in equation (4.1) by means of physical invariants such as the kinetic and potential energies. For example, in most problems it is possible to write

$$-\sum^{k} \mathbf{F}_k \,.\, \delta\mathbf{r}_k = \delta V - \sum^{j} \mathscr{F}_j \,.\, \delta\mathbf{r}_j. \tag{4.2}$$

where V is a potential function of the coordinates, while $\mathscr{F}_j$ are given 'applied forces'. Equation (4.1) may then be written as

$$\sum^{k} m_k \ddot{\mathbf{r}}_k \,.\, \delta\mathbf{r}_k + \delta V = \sum^{j} \mathscr{F}_j \,.\, \delta\mathbf{r}_j. \tag{4.3}$$

The extension of this result to continuous systems is obtained without difficulty if we consider the physical situation from the standpoint of the physicist. In that case we may describe the displacement field by writing it in the form

$$\mathbf{r} = \mathbf{r}(q_1, q_2,..., q_n, x, y, z, t), \tag{4.3 a}$$

where x, y, z are the initial coordinates, t is the time, and $q_1, q_2,..., q_n$ are generalized coordinates. Since a physical system is essentially of molecular or atomic composition we may divide it into a large number of cells. Each cell should contain enough molecules or atoms for the macroscopic laws to be valid and at the same time be small enough, so that their behaviour may be described by a few parameters as discrete entities analogous to particles.

In the context of thermodynamics, similar concepts were also discussed in Section 3 of Chapter 1. The physical completeness of the representation (4.3 a) of the field $\mathbf{r}$ by the generalized coordinates is evident if we divide the continuous system into N cells and choose a non-singular transformation of the $3N = n$ components of displacements of the centre point of each cell into the n variables q_i. For the continuous systems, the summations of equation (4.3) are then replaced by integrals. We obtain

$$\iiint_{\tau} m\ddot{\mathbf{r}} \,.\, \delta\mathbf{r}\, d\tau + \delta V = \iint_{A} \mathscr{F} \,.\, \delta\mathbf{r}\, dA. \tag{4.4}$$

We have assumed here, for simplicity, that the forces $\mathscr{F}$ are applied to the boundary A of the domain of volume integration τ.

As can be seen, we may consider this description of the continuous displacement field $\mathbf{r}$ by a finite number of coordinates q_i as introducing a *constraint*, which limits the configuration of the field $\mathbf{r}$ only to those that are physically significant. In the terminology of optics, one could say that the physical description does not have to extend below a certain *resolution threshold* beyond which the continuous model represented by the differential equations loses its physical significance.

The variational principle (4.4) leads directly to the Lagrangian equations,

$$\frac{d}{dt}\left(\frac{\partial \mathscr{T}}{\partial \dot{q}_i}\right)-\frac{\partial \mathscr{T}}{\partial q_i}+\frac{\partial V}{\partial q_i} = Q_i. \tag{4.5}$$

This result is based on an important property of the term containing the acceleration in equation (4.4), namely, that it may be expressed by means of a physical invariant $\mathscr{T}$, which represents the kinetic energy. The derivation depends on the functional identity

$$\frac{\partial \mathbf{r}}{\partial q_i} = \frac{\partial \dot{\mathbf{r}}}{\partial \dot{q}_i}. \tag{4.6}$$

The foregoing procedure is essentially the same as followed in the derivation of the Lagrangian equations for heat conduction. For example, in the case of isotropic properties the derivation is based on the variational principle (2.9) of Chapter 1, which is written

$$\iiint\limits_{\tau} \frac{1}{k}\dot{\mathbf{H}}\,.\,\delta\mathbf{H}\,d\tau+\delta V = -\iint\limits_{A} \theta\mathbf{n}\,.\,\delta\mathbf{H}\,dA \tag{4.7}$$

and is analogous to equation (4.4). The resulting Lagrangian equations,

$$\frac{\partial V}{\partial q_i}+\frac{\partial D}{\partial \dot{q}_i} = Q_i, \tag{4.8}$$

follow from the possibility of expressing the term containing $\dot{\mathbf{H}}$ by means of a physical invariant, which in this case is the dissipation function D. Again this requires the use of the functional identity

$$\frac{\partial \mathbf{H}}{\partial q_i} = \frac{\partial \dot{\mathbf{H}}}{\partial \dot{q}_i} \tag{4.9}$$

similar to relation (4.6).

Actually, a large category of phenomena are governed by Lagrangian equations that involve both the kinetic energy and the dissipation function. Their general form is

$$\frac{d}{dt}\left(\frac{\partial \mathscr{T}}{\partial \dot{q}_i}\right)-\frac{\partial \mathscr{T}}{\partial q_i}+\frac{\partial D}{\partial \dot{q}_i}+\frac{\partial V}{\partial q_i} = Q_i. \tag{4.10}$$

This result, for dissipation due to viscous friction, has been known for a long time since the introduction of Rayleigh's dissipation function. That such a unified treatment is also possible for the case of thermodynamic dissipation has been shown by the author. This is illustrated for the particular case of thermoelasticity in the preceding section.

Approximate and weak solutions

In the foregoing discussion we have considered the generalized coordinates as a complete representation of the physical system by means of a large but finite number of variables. However, it is important to point out that the most useful applications of the Lagrangian equations do not require that the generalized coordinates provide a complete physical description. In many cases it is possible to guess that the solution falls within a *family of functions with a small number of parameters*. These parameters may then be chosen as generalized coordinates obeying Lagrangian equations. As already indicated in Section 4 of Chapter 1, in the context of heat transfer this procedure provides one of the most powerful methods of obtaining approximate solutions by taking advantage of known features of such solutions in the formulation of the problem, thus providing a drastic reduction in the number of unknowns.

At the other extreme, generalized coordinates also provide solutions of a type known mathematically as *weak solutions*. This will occur when using an infinite but denumerable set of generalized coordinates to describe the physical system. An example of this was provided in Section 7 of Chapter 2, where the infinite set of generalized coordinates is represented by unknown coefficients in an infinite series. Such weak solutions may not be solutions of the differential system at certain points finite or infinite in number. The total measure of this set of points is zero. However, the solution may still be said to represent an infinite resolution of the physical system in terms of an arbitrarily large number of cells. These features do not restrict the physical completeness of the solution since the resolution goes much beyond the physical limitations of the continuous model.

Virtual work as a variational scalar product

The concept of virtual work has long been known in the context of mechanics as providing a powerful tool in the analysis of complex problems and the formulation of simplified approximate equations. As shown by the author, the concept can be extended to thermodynamics. Actually, the virtual work concept is nothing but the expression in a physical context of a notion that may be translated into the language of functional analysis.

Consider, for example, an unknown scalar field

$$\varphi = \varphi(x_1, x_2, \ldots, x_n, t) \tag{4.11}$$

function of $n+1$ variables x_i and t. Let us assume that it is governed by the equation

$$\mathscr{P}(\varphi, x_i, t) = 0. \tag{4.12}$$

This relation may represent a differential equation of a very general type. Let us multiply equation (4.12) by arbitrary variations $\delta\varphi$ and integrate the result over a volume τ in the n-space of the variables x_i. We obtain

$$\int_\tau \mathscr{P}\,\delta\varphi\,d\tau = 0, \tag{4.13}$$

where

$$d\tau = dx_1\,dx_2...\,dx_n.$$

For arbitrary variations, $\delta\varphi$ equation (4.13) does not guarantee that equation (4.12) is verified everywhere. However, in mathematical terminology it is said to be verified *almost everywhere*. By this is meant that the set of points at which it is not verified has a zero measure.

A geometric interpretation of equation (4.13) is obtained by putting

$$\mathscr{P} = \epsilon. \tag{4.14}$$

This value ϵ of $\mathscr{P}$ may be considered as a vector in functional space. Equation (4.13) is written

$$\int_\tau \epsilon\,\delta\varphi\,d\tau = 0. \tag{4.15}$$

The integral may be considered as the scalar product in functional space of the two vectors ϵ and $\delta\varphi$. The value of ϵ also represents the 'error' of equation (4.12). Hence, in the language of functional space, we interpret equation (4.14) by saying that the projection of ϵ on a vector $\delta\varphi$ vanishes for all orientations of this vector.

Formulation of physical problems by variational relations such as equation (4.13) opens the way to the use of generalized coordinates and the derivation of the corresponding Lagrangian-type equations. As amply demonstrated by the applications, the equations may be usefully transformed by operations such as integration by parts and by using other transformations, so as to bring out groups of terms that correspond to invariants. The remaining terms play a role analogous to the generalized forces.

A number of obvious generalizations of the foregoing procedures is immediately evident. For example, instead of equation (4.13) we may write

$$\int_\tau \mathscr{P}\,\delta\mathscr{F}\,d\tau = 0, \tag{4.16}$$

where $\mathscr{F}$ is a function of φ. More generally, it may even be a differential expression containing φ. Another generalization is obtained by noting that the number of independent variables need not be $n+1$, but may be $n+2$, $n+3$, etc. For example, consider an unknown field

$$\varphi = \varphi(x_1, x_2,..., x_n, t_1, t_2) \tag{4.17}$$

function of $n+2$ variables. We may also write a variational equation such as (4.15). However, in this case the generalized coordinates will be functions of two variables, t_1 and t_2. The generalized Lagrangian type equations will then be partial differential equations.

The method has been illustrated here in terms of a scalar field φ but it is obviously applicable to vectors and systems of equations with several unknowns, as shown in the case of mechanics where the unknown is a displacement field with three components.

Transformation in functional space

We may look upon the variational scalar product as an invariant that provides a method of great power and flexibility for transformations of equations in functional space. A typical case is brought out by the problem of heat conduction treated in Section 7 of Chapter 1, where the unknown temperature distribution is transformed into an unknown penetration depth. This constitutes a radical departure from the usual type of functional transformations, since it substitutes for the unknown field an unknown abscissa corresponding to prescribed values of the field. Such powerful transformation methods are applicable to wide areas of physics.

Ignorable subspace

The concept of ignorable coordinates leads to the more general one of ignorable subspace. This was discussed in Section 6 of Chapter 4 in the context of heat transfer. The ignorable subspace may be considered from the standpoint of transformations in functional space. It corresponds to a partition of the space into separate subspaces, which are uncoupled. In each subspace the solutions will satisfy a different type of equation. This procedure is applicable to a large category of problems in mathematical physics. For example, in addition to its application to heat conduction considered in Chapter 4, it may be applied in acoustics.‡ In this case the ignorable subspace is the divergence-free incompressible flow which satisfies Laplace's equation.

‡ For a Lagrangian treatment of acoustics, see the paper by M. A. Biot and I. Tolstoy, 'Formulation of wave propagation in infinite media by normal coordinates with an application to diffraction', *J. acoust. Soc. Am.* **29**, 381–91 (1957).

Method of minimum squares

A well-known procedure of obtaining approximate solutions of differential equations is to minimize the integrated square of the error ϵ. For example, in the case of the differential equation (4.12) this integral is

$$I = \int_\tau \epsilon^2 \, d\tau = \int_\tau \mathscr{P}^2 \, d\tau. \tag{4.18}$$

A solution that minimizes this expression satisfies the equation

$$\delta I = \int_\tau \mathscr{P} \, \delta\mathscr{P} \, d\tau = 0. \tag{4.19}$$

As can be seen, this is a particular case of equation (4.16) obtained by choosing $\mathscr{F} = \mathscr{P}$. Hence the method of minimum squares may be considered as corresponding to a particular choice of the variational scalar product. However, since $\mathscr{F}$ in this case is not a function but an operator, the differential equations obtained by minimizing the value of I are usually more complicated and of a higher order than the original equation (4.12). Furthermore, the fundamental physical invariants of the problem are generally lost in a formalism that is artificial and remote from the physical problem.

Galerkin's method

This method may also be considered as a special case of the general procedure based on the variational scalar product. It is obtained by choosing a linear dependence on the generalized coordinates and treating the latter as unknown constants. For example, in the case of the differential equation (4.12) we write the solution in the form

$$\varphi = \sum^k q_k \, \varphi_k(x_i, t), \tag{4.20}$$

where $\varphi_k(x_i, t)$ are suitably chosen functions of the $n+1$ variables x_i and t. We then write the integral

$$\int_{\tau'} \mathscr{P} \, \delta\varphi \, d\tau' = 0, \tag{4.21}$$

where
$$d\tau' = dx_1 \, dx_2 \ldots dx_n \, dt. \tag{4.22}$$

Hence, in this case the domain of integration includes all the $n+1$ variables x_i and t, instead of only the n variables x_i as in equation (4.13). By substituting the value (4.20) of φ into equation (4.21) we derive

$$\sum^k \int_{\tau'} \mathscr{P} \varphi_k \, \delta q_k \, d\tau' = 0. \tag{4.23}$$

Since the variations δq_k are arbitrary, this leads to the equations

$$\int_{\tau'} \mathscr{P} \varphi_k \, d\tau' = 0, \tag{4.24}$$

which express the orthogonality between the error $\mathscr{P} = \epsilon$ and the functions φ_k. These equations embody Galerkin's method.

Hence the method is derived from the general equation (4.16), first by putting $\mathscr{F} = \varphi$, secondly by assuming a linear dependence of φ on the generalized coordinates, treated as constants, and thirdly by integrating in a domain that includes all the variables. This last feature amounts also to saying that the variable t does not appear explicitly while there are $n+1$ variables x_i.

The restricted flexibility of Galerkin's method is due partly to the imposed linear dependence of the unknowns on the generalized coordinates and partly to the fact that these coordinates are assumed constant. This is in contrast with the Lagrangian formulation where the generalized coordinates are of a very general type and obey simplified differential equations. In many problems the dependence on some category of variables such as space coordinates is easily approximated while the dependence on other variables such as the time remains to be determined. This dependence is easily analysed as a separate problem by the use of Lagrangian-type equations.

The Galerkin method may, of course, be used to solve approximately the Lagrangian equations themselves. Such a two-step procedure may be advantageous when the Lagrangian equations provide a clue to suitable approximations of generalized coordinates as functions of time.

AUTHOR INDEX

SUBJECT INDEX

PRINTED IN GREAT BRITAIN
AT THE UNIVERSITY PRESS, OXFORD
BY VIVIAN RIDLER
PRINTER TO THE UNIVERSITY